T0255355

Stories from the Deep Earth

Geoffrey F. Davies

Stories from the Deep Earth

How Scientists Figured Out What Drives Tectonic Plates and Mountain Building

Geoffrey F. Davies
Research School of Earth Sciences
Australian National University
Canberra, ACT, Australia

ISBN 978-3-030-91361-8 ISBN 978-3-030-91359-5 (eBook)
https://doi.org/10.1007/978-3-030-91359-5

This Springer imprint is published by the registered company Springer Nature Switzerland AG
The registered company address is: Gewerbestrasse 11, 6330 Cham, Switzerland

Contents

Chapter 1
Deep Earth and Deep Time, Big Ideas and Big Egos

I was preparing dinner, in one of my alone periods, my brain idling as usual on my current preoccupations about how the deep interior of the Earth might work. I probably looked very absent-minded. Perhaps I was.

Standing thus in the middle of the kitchen, it struck me that the hybrid rock in the mantle would also sink, just like the former oceanic crust. Both would tend to collect at the bottom. Mantle plumes would suck up the mixture of unlike materials. That would explain the contradictory signatures in rocks erupted at Hawaii and other places. One material, the old oceanic crust, would record a history of being melted, changed and erupted. The other, the hybrid, would retain traces of 'noble' gases (helium, neon, argon) surviving from the long-ago formation of the Earth.

The great contradiction—a source of material from the deep Earth seemingly both highly processed and unprocessed, or 'primitive'—would be explained. And it would be explained in a natural way in a scenario I had already developed for other reasons. Wow! Perhaps there was a way to reconcile *all* of the geochemical evidence, *all* of the geophysical evidence, and the physics of very slow flow of very stiff rocks deep in the Earth.

The jargon in that little anecdote may not make any sense to you, but it will be explained in due course. The point here is that in that moment in 2009 I thought I had seen a way to reconcile *all* of the evidence regarding how the deep interior of the Earth behaves. That would be no mean milestone, because for about forty years great debates had been raging. The geophysicists disagreed with the geochemists. The geophysicists disagreed with each other. The geochemists disagreed with each other. You could be forgiven for thinking we were talking about *four* planets, each quite different from the other. We were like the committee of blind men feeling the parts of the elephant. But we were all looking at the same planet, the one we live on. Really, wouldn't it be better if we could find *one* version of the planet that accommodated *all* of the evidence? As we say these days, regarding global warming, there is no Planet B.

We were arguing because it was not at all obvious how to reconcile all of the evidence within one picture. There was more than one seeming contradiction among

G. F. Davies, *Stories from the Deep Earth*,
https://doi.org/10.1007/978-3-030-91359-5_1

the various lines of evidence. Another difficulty was that few people had made the effort to gain a deeper understanding of all the evidence, the geophysics and the geochemistry. We were educated in one discipline or the other, and the work was highly specialised. You understood your own work, you got your results and drew your conclusions. If the other side disagreed with you they must be wrong.

Well that is one attitude, and it was not uncommonly expressed, sometimes in forceful terms. Another view might be that if our evidence seems to conflict then there is something we don't understand. If that's true then there is more still to be learned. Perhaps there are unrecognised biasses or errors in some measurements. Perhaps some of the computer models are too simplified, or are built on shaky assumptions. Perhaps some of the common inferences, taken for granted because they've been around for decades, are not actually justified. If you wanted to understand how the Earth really worked, perhaps you needed to work carefully through all those possibilities.

The trouble is there are pressures on scientists these days to keep churning out results, so they can get their next grant of research money, so they can support their students to do the work, so they can publish papers and justify their next grant. It is easier to keep doing what you're doing. It would be risky to back off and really explore all the possibilities because you might not discover anything and your rate of publication of papers might fall away, your grant money dry up and your career go into a tailspin.

And then there is the small matter of egos. If I am right, then some of you are wrong, and vice versa. Supposedly, it is laudable to propose an idea that stimulates useful research but turns out ultimately to be not the best explanation. However most of us want to be the ones who get to the 'right' explanation, the explanation that becomes generally accepted. Some of us want it more than others. Some of us are desperate to be right, and will go to great lengths to try to discredit others.

You thought science was a cool and dispassionate collecting of evidence, leading inexorably to the Truth? Sorry, wrong on several counts. There are big gaps in the evidence, and what we have is confusing. People come up with incompatible, even outlandish, hypotheses to explain the evidence. Science is not actually about Truth, or even truth, though that claim might shock some people. We scientists all have skin in the game, as they say. Some fields are more acrimonious than others. Some are even fairly polite. Figuring out the inside of the Earth tended towards the acrimonious end of the spectrum.

Thomas Edison is reputed to have said that genius is one percent inspiration and ninety nine percent perspiration. You might have a bright idea, but there's usually a lot of work to establish that it's reasonable, and more work to persuade others that your evidence is compelling. So it was with my bright idea about hybrid rock sinking through the mantle and resolving one of the remaining big puzzles. There would be more work over the next year or two to develop the idea, to talk about it at conferences, and to try to get it published. It might be quickly recognised as significant, or only slowly. Someone might point out some contrary evidence or other reasons why it would not work. You put it out and wait for the response, ready to debate the pros and cons, perhaps for some time.

What I was not expecting was that it would be greeted by a resounding silence. I reported it at three conferences, published it in a prominent journal and included it in a review paper and a book chapter. It was then pretty much ignored, for about a decade so far.

I want to tell a story of which that anecdote is just one small part. The bigger story is about how the inside of the Earth overturns, very very slowly, and how the overturning brings new material from great depths right to the surface, and it takes other material from the surface and sinks it hundreds, thousands of kilometres back into the depths. Between the rising and the sinking, pieces of the crust move slowly across the Earth's surface. The crust is broken into a dozen or so gigantic pieces, like a jigsaw puzzle, only the pieces are moving: pulling apart here, thrusting together there, sliding past each other somewhere else. The movement of these pieces is called *plate tectonics*, which has become fairly well known. The operation of plate tectonics has profound consequences for us, living on the Earth's surface, for example by rejuvenating landscapes and soils, plausibly helping to drive the proliferation of life, and even just keeping the continental surfaces above sea level. How did we arrive at this far-from-obvious and perhaps unsettling view of the planet beneath our feet?

I also want to tell a story of how such science has been done. Not just who did what, but why? In what context did they operate and think, what were their strengths and vulnerabilities, and what, precisely has been accomplished? Scientists are people too, and they bring all the strengths and foibles of people into their practice of science. Science is both venerated and derided in our society, but what level of esteem is really appropriate? Some of the story of doing the science will be told as historical anecdotes and the later parts will be threaded around my own experience. The latter parts do get a bit detailed at times, but you don't need any technical background and I explain the relevance of the evidence to the larger story as we go along.

The latter part will also necessarily involve a point of view, my own best judgement of situations and debates that may still be contested. I try to give a fair account, but inevitably others would weigh things differently. I think it's better to give a personal account, with this forewarning, than to make a pretence of 'balance'. I also include some non-scientific aspects of encounters with peers and with the system, which at times dished out some treatment that was rather shabby in my view. My experience will be far from unique, and certainly not the worst, but that is the larger point: if we don't bring such stories to light then poor behaviour will continue undiminished. There are many misunderstandings and myths about science and its practice and it is healthy for our society to be given more insight into the kinds of debates and encounters that actually go on.

To be clear, this is not a professional history. It is based on my personal experience, and on reading that is necessarily very wide, because the subject relates to very diverse aspects of Earth science. Nor is it a professional review. I use material to make key points in the story, with no attempt to survey all the work going on at any given time. Where others' work is pertinent it is used, but my own work naturally features a lot in the story I was pursuing, which is the story told here.

Coming back to the Earth story, it took a long time for someone even to recognise the existence of those great crustal pieces, the *plates*, and to realise they are the reason for most earthquakes, volcanoes, and mountain building. One reason it took a long time is that no-one had seen anything else quite like them, there was no evident analogue to give people a clue. Another reason is that the clearest evidence is under the sea, and the sea floor was only explored in sufficient detail after World War II. Another reason again is that most geologists, who were gathering the evidence, don't know much about fluid motion. That is not a criticism. We depend on many specialists gathering a lot of detailed information, but it may take people with other kinds of expertise to see how the details fit together.

Once the plates were recognised, only in 1965, there was a long debate about what was driving the movement of the plates, and indeed why are there plates at all? That eventually became my own major interest. I think we got the physics of the motion worked out pretty well, after about thirty years of arguing. In the meantime geochemists, who measure the abundances of trace elements and isotopes in rocks and use them to fingerprint various kinds of rocks and processes, developed their own story, but their story contradicted the story developed by geophysicists, or at least one of the geophysical stories—that's where my opening anecdote fits in.

There is another part of the story, concerning so-called mantle plumes. They are believed to be the source of volcanic eruptions at Hawaii, Iceland and a dozen or more other places. They are distinct from the plates and their place in the story also involved long debates.

So the story of plates and plumes spans over fifty years, so far, but really it is the continuation of a story stretching back two or three centuries. The story really starts as the science of geology was coming into being, because a few people were trying to assemble and interpret the common observations of mountains and plains, layered rocks that might be flat, or tilted on edge, or folded in great curves, and other rocks that were not layered at all but looked as if they crystallised in place, perhaps from solution in water or perhaps from having been melted, molten.

Geology gave rise to one of the really big shifts in human perception, less remarked than the shift to thinking of the Earth as a spinning sphere travelling a wide path around the sun. Geology's big shift was the realisation of *deep time*. That involved a rancorous debate between a new breed of geologists and those, geologists and others, who held to the Biblical account of our world being a few thousand years old. This was a particularly vexed issue in the West; some other cultures did not have such a foreshortened view of history. Not everyone regards that debate as settled, though for the vast majority of scientists deep time is the accepted interpretation, intrinsic to modern astronomy as well as geology.

One scientist, encountering the evidence for deep time with a geologist friend in 1788, said 'the mind seemed to grow giddy by looking so far into the abyss of time'. Well might our minds grow 'giddy', because it soon became clear that deep time involved not just millions of years but hundreds of millions of years. Then it gradually emerged that the rock record goes back even much further than that, through *billions* of years. The current estimate for the age of the Earth is a little over 4.5 billion years. In the meantime astronomers have decided that the universe is about 13 billion years old. No-one can conceive of a million years in any subjective sense, let alone a billion years.

So Copernicus' shift to putting the sun at the centre of the solar system led us to the notion that the universe is unimaginably vast. Geologists needed enough time for very slow processes to accumulate into what they were observing, and that led to the notion that the Earth and the universe are unimaginably old.

The conception of deep time led to another great shift: that the incredible diversity of living forms on the Earth evolved through the very gradual accumulation of small variations in organisms. Indeed Charles Darwin was directly involved in the debates about deep time, because he recognised the need for vast amounts of time for his theory of natural selection to work its wonder, just as the geologists needed time for their slow processes of deformation and erosion.

There is one more part to the story of the deep, creeping interior of the Earth, perhaps more mundane than grand theories of an ancient universe. It is the story of how heat affects the properties of solids and liquids, and to a lesser extent how intense pressure affects them. This will be familiar to a degree, but the implication for our story is probably not so familiar. The more familiar part is that if you heat a piece of iron it will soften so it can be bent into various shapes, such as a horseshoe. The less familiar part is that this can also be true of rocks. If rocks are heated enough, and they are confined by sufficiently high pressure, they also will deform into a different shape without breaking. They will deform rather more slowly than a piece of iron, but we have just indicated that in this business we have plenty of time at our disposal, so even something going very slowly can get a long way, given enough time.

The realisation that 'solid' rock might slowly deform had its own checkered history. A few people, mid-to-late nineteenth century, put together observations of the Earth and got a fairly clear picture. Yet the insight remained neglected. This is perhaps a little strange, because geologists could see rock strata folded into great curves, and the implication is that they had been deformed rather than broken by whatever great forces had acted on them. Perhaps many geologists were not so interested in the physical process of deformation, they just needed to know it had happened. On the other hand when the idea of continental drift emerged it was disparaged by physicists and mathematicians as being impossible because the rocks of the mantle were known by then to be solid.

The catch was that neither geologists nor physicists necessarily had much knowledge of how heat and pressure affect the mechanical properties of solids. That is the domain of materials scientists, including metallurgists. There must have been a lot of long-standing knowledge of metals and ceramics, because they have played large parts in our civilisation, but perhaps they were regarded as crafts, rather than proper quantitative sciences. It would not be the only example of scientists taking too narrow a view of where knowledge is to be found.

Marcia Bjornerud, in her 2020 book *Timefulness*, suggests geology itself may be a bit of an orphan science. As she says, it has no Nobel prize and is not always taught as a basic science subject. Certainly I have encountered physicists, not well informed, who seem to think it is a derivative activity involving banging on rocks with a hammer and collecting and cataloguing things, a bit like stamp collecting. In fact geology, in the broadest sense of studying the 'solid' Earth, uses a wide range of sophisticated instruments and methods appropriated from other sciences, plus

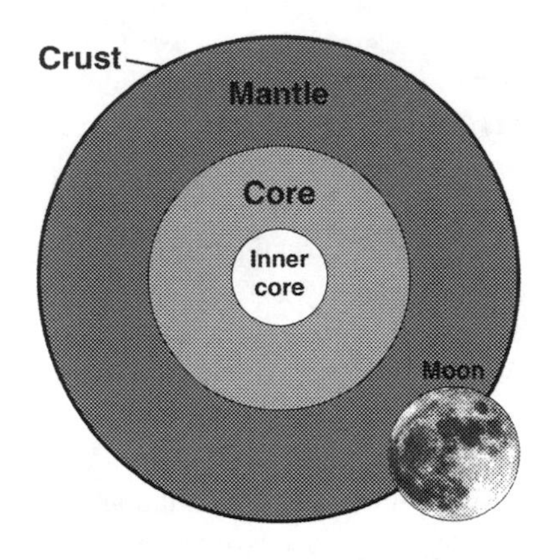

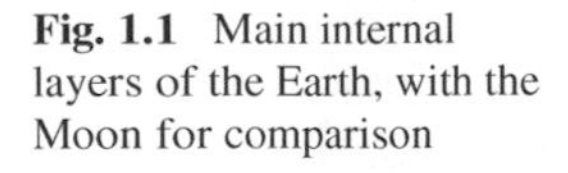
Fig. 1.1 Main internal layers of the Earth, with the Moon for comparison

many more that are specific to the subject. But Bjornerud gives us a more nuanced description of geology, as she recounts her surprise, on first encounter, that it

> ... applied scholarly habits one associates with the study of literature and the arts—the practice of close reading, sensitivity to allusion and analogy, capacity for spatial visualisation—to the examination of rocks. Its particular form of inferential logic demanded mental versatility and a vigorous but disciplined imagination. And its explanatory power was vast; it was nothing less than the etymology of the world.

Yes, a perceptive description. I would just add that it requires an ability to sift through large amounts of confusing information looking for a nugget, an observation that implies something important and is not likely to be quickly superseded.

So our story is of deep time, deforming rocks, mountain building. It is a story of how an understanding slowly emerged that the Earth produces great forces that slowly move the surface, and the interior, in ways that produce what we see around us on our amazing planet. The story is also about the people who figured these things out, and how and why they did what they did.

A tradesman once asked me which university department I worked in. I said 'Earth and Planetary Science', but I must have mumbled. He thought I said 'Earth and Plant Science'. So to be sure you don't think this is a book about gardening or horticulture I am capitalising the name of planet Earth. The other kind of earth is where earthworms live and plants grow.

Let us get our story located. We're talking about the inside of the Earth. Most people will have heard that the inside of the Earth is hot, and that it has a molten core. However the molten core is not the source of the lava that flows out of volcanoes, as newspapers often mistakenly say, because the molten core is far deeper and made mostly of iron. The source of volcanic lava is a lot shallower and a little more subtle to understand. We'll get to that in due course.

The Earth's interior is divided into three main layers, which are illustrated in Fig. 1.1. The outermost layer is the *crust*, which is quite thin compared with the size of the Earth, varying from around 7 km (7 kms) thickness under the sea floor to about 50 km under mountains. The crust is made of rock of varying composition. We don't have to worry too much about the details of rock compositions, but it is useful to mention that basalt is a dark, fine-grained and moderately dense volcanic rock, and granite is a less dense and lighter-coloured rock, commonly with large crystalline grains. The oceanic crust is made mainly of basalt. Most of the continents are made of rocks that vary from something between basalt and granite to, in places, pure granite.

The crust is so thin compared with the size of the Earth that it hardly shows in a diagram like this one. Earth's diameter is 12,740 km. The Moon is included in Fig. 1.1 for comparison; its diameter is 3480 km, a little over a quarter of the Earth's diameter.

Next below the thin crust is the *mantle*, which extends to a depth of about 2900 km, almost half way to the centre of the Earth. The mantle is made of solid rock, but we'll see later it is not quite as solid as we would normally think solid rock to be. Anyway the mantle is a little denser than the crust, and its main constituent is *olivine*, which is a dark green semi-precious stone. Well, the mantle is mostly olivine near the top, but deeper down the extreme pressures transform its crystal structure into denser forms. We'll get to that later too.

Below the mantle is the core, and most of it *is* molten, so finally we have arrived at the 'molten core'. The core is made mostly of iron, alloyed with some lighter elements that might be silicon or sulphur and other minor components. You might wonder how we know the core is made mostly of iron, and strictly speaking we don't. The core is *inferred* to be mostly iron because iron is a common heavy element in the sun and in meteorites, and the density of the core is close to what the density of iron would be at the very high pressures and temperatures inside the Earth. I haven't yet told you how we know the density of the core, nor how we know it is molten, but the evidence for those claims is much more direct, as I'll explain in a while.

Here we already have an excellent illustration of how studying the inside of the Earth keeps you humble, or ought to. Our knowledge of the core is indirect. Its composition is reasonably inferred as I just explained, but it is not certain. No-one has ever been down to the core, not even Jules Verne (who wrote *Journey to the Centre of the Earth*), and no-one is ever likely to. It is more inaccessible than the surface of Neptune. No-one has ever been down to the mantle either, but bits of the mantle are brought to the Earth's surface during mountain building, so we have a more secure grasp of its composition. On the other hand no bit of the core has ever been known to have been brought to the Earth's surface, so we have to resort to *inference*.

This business of what we 'know', scientifically, how we know it, and how secure is that knowledge, is profoundly important for our modern 'scientific' society. Just think of the arguments about global warming, evolution, and even whether a virus causes illness. Our societies would be in less conflict if more of us understood that scientific 'knowledge' is not immutable Truth, but neither is it just opinion.

Anyway, you can see in Fig. 1.1 that the core is rather bigger than the Moon and, at 6940 km diameter, even slightly bigger than Mars (6770 km). Within the core is the 'inner core', which is solid, not molten, and slightly denser than the 'outer' core. Current thinking is that when the Earth started out it was rather hotter inside and has been cooling very slowly ever since. At first the core was completely molten, but eventually the molten core material started solidifying and collecting at the centre, thus forming the inner core.

There is vigorous debate at the moment about whether the inner core started forming several billion years ago, relatively 'early' in the Earth's history, or 'only' about one billion years ago. You might think this question is about as important as how many angels can dance on the head of a pin, but it is all tangled up with questions like how the Earth's magnetic field is generated and how strong are the upwelling mantle plumes that we will meet later. The magnetic field protects us from dangerous cosmic rays, and the Hawaiian and Icelandic volcanoes are believed to be generated by mantle plumes, so this question has some relevance to people's lives.

You might have noticed that, as you go down, each layer is denser than the one above. This is what you would expect if the Earth's interior was once hot enough to flow, so that denser material would sink to the bottom and lighter material would float to the top. It is pretty clear by now that planets form from material left over from star formation. Rocky and icy material left over from forming the sun slowly collected into large bodies. As the bodies became planet-sized, each collision of new material falling on them released a lot of energy, enough to melt everything by late in the process. So there is good reason to think the Earth did start hot and its interior has been cooling, very slowly, ever since.

This completes our first tour inside the Earth. There is much more detail to explore, but at least by now you'll have more idea what, and where, I'm talking about if I get excited about mantle convection.

How do we know the Earth is layered like this? It is through of *seismic waves*, which are sound waves generated inside the Earth by earthquakes. The seismic waves travel down through the Earth and reflect off the boundaries between the layers. This is illustrated in Fig. 1.2b, which shows some typical paths followed by seismic waves. The paths through the mantle are curved because the waves travel faster the deeper they go, which results in them being *refracted*.

Seismologists have long-since placed many seismometers all around the Earth, and they record the seismic waves from earthquakes all around the Earth as the waves arrive back at the surface. Figure 1.2c shows a typical seismogram recorded at one place as various waves arrive, some reflected, some refracted, some travelling around the Earth's surface like ocean waves. Since late in the nineteenth century seismologists have been recording the 'travel time' of the various seismic waves *versus* distance from the earthquake. They can then work backwards to figure out how deep the waves went and whether they have been reflected or refracted. They can then calculate the positions of the internal boundaries. The accuracy and detail of these calculations have been refined for over a century. The result is that seismic waves give us the most detailed and accurate picture of the Earth's interior, rather like a medical ultrasound.

Figure 1.2a illustrates how an earthquake generates seismic waves. An earthquake occurs when two parts of the crust suddenly slip past each other along a *fault*. The crust in some active earthquake zones is being very slowly deformed all the time. Most of the time friction prevents any slippage on the fault, and instead the rocks bend (by a very small amount), but eventually the stress overcomes the friction and

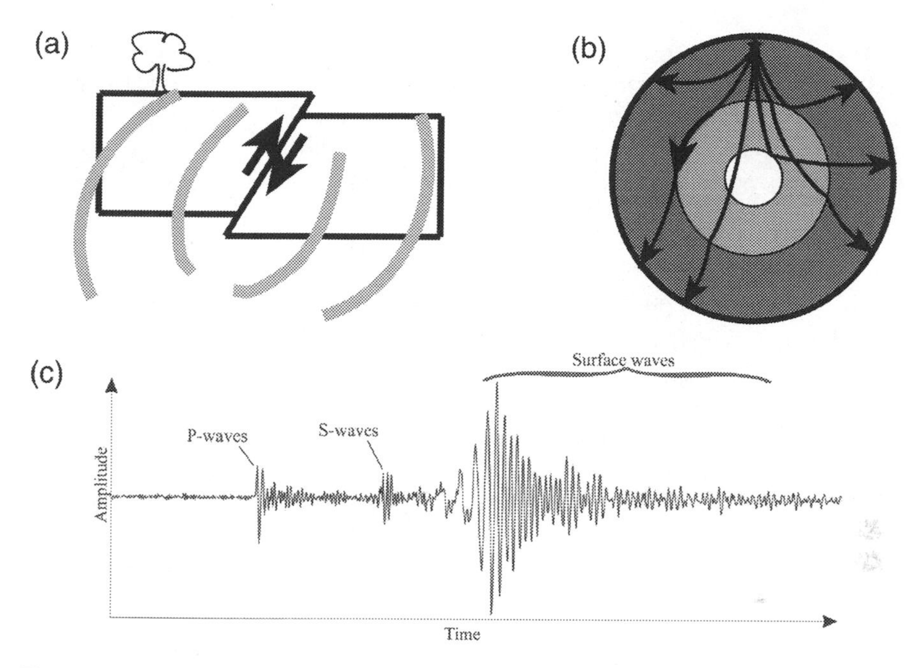

Fig. 1.2 **a** An earthquake occurs when slowly building stress overcomes friction on a fault and the blocks of crust on either side of the fault suddenly slip past each other. The sudden motion sends elastic waves (grey bands) travelling out through the crust and into the Earth's interior. **b** Typical paths followed by seismic waves inside the Earth, reflecting off boundaries, refracting through boundaries and diffracting around the core. **c** A seismogram, which is a record of the waves arriving from a distant earthquake. Different kinds of waves travel at different speeds and arrive at different times (Brian Ricketts, creative commons [1])

the two sides of a fault suddenly slip past each other. It is this sudden slipping that sends sound waves out through the inside of the Earth. As the waves reach the surface they shake the ground, and that shaking is what people experience as an earthquake.

Already we are encountering some unfamiliar properties of rocks. You may not be used to thinking of rock as being *elastic*, but a force on it will cause it to bend very slightly. If the force becomes too great the rock will break. That is what you see if you hit a rock with a hammer—it is brittle and a piece breaks off. However that tiny bit of elastic bending before it breaks is enough to allow elastic waves to travel out from an earthquake. Seismic waves are this kind of elastic wave.

Another result of the slight elasticity of rocks is that they compress under pressure. As you go down into the Earth, so to speak, pressure increases rapidly because of the weight of rock above. Now it happens that the seismic wave *speed* tells us about the *compressibility* of the rock the wave is passing through. This allowed clever seismologists to calculate their way down from the surface, figuring how much the rock compresses, and the density increases, with each step down. Rock at the bottom of the mantle is compressed about 30% by volume compared with rock at zero pressure—not really very much given the enormous pressure at that depth.

There are a couple of complications in calculating the density *versus* depth. The crystal structures of the rocks re-arrange their atoms into denser forms a few hundred kilometres down, which increases the rock density. Also the density takes a sudden jump up as you go into the iron core. Fortunately there is other information (the Earth's total mass and its moment of inertia) that allows the core density to be constrained. If this doesn't make a lot of sense don't worry, I just want to give you the idea that the density of the interior is quite well constrained from real information, and not just guessed at.

There is one more thing I haven't explained, and that is how we know whether a layer is liquid or solid. It is because solids transmit *two* kinds of elastic waves, whereas liquids and gases transmit only one kind. Sound waves in the air alternately compress and expand the air as they pass. So do sound waves in water, even though water is much less compressible than air. Solids also transmit an elastic wave that alternately compresses and expands the solid (very slightly), but solids transmit a *shear* wave as well. If you and a friend hold a skipping rope loosely between you and you wiggle your end of the rope sideways you should see waves of 'sideways' motion travelling along the rope. A shear wave in a solid jiggles the solid sideways as it passes, whereas a *compressional* wave jiggles the solid back and forth in the direction the wave is travelling.

A shear wave only changes the shape of the solid, whereas a compressional wave expands and contracts the solid. A liquid does not resist shearing, you can push it into any shape you want, so a liquid will not transmit a shear wave. On the other hand a solid does resist changing its shape, and it will spring back to its original shape if you stop applying a force to it. This is why a solid can transmit a shear wave.

Now to the point. Earthquakes generate both compressional waves and shear waves. The shear waves travel more slowly, so they are easy to distinguish in a seismogram like the one in Fig. 1.2c. Both compressional and shear waves can travel through the crust and mantle, so we conclude that the crust and mantle are solid. Compressional waves can also travel right through the core and come out on the other side of the Earth, but there are no shear waves coming out on the other side of the Earth. The core makes a shadow zone in which no shear waves arrive. We conclude that the core is liquid, and unable to transmit shear waves.

So we learn that the crust is solid, the mantle is solid and the outer core is liquid. The arguments are a little more complicated and the observations more difficult, but it was eventually shown that the inner core is solid.

That establishes the setting for our story, which is mostly about the mantle. You might think if the mantle is solid then there is not much story to tell. It will just sit there, being solid and doing nothing except transmit the odd seismic wave. That indeed was the conclusion of many scientists from late in the nineteenth century, when the solidity of the mantle was established, until the middle of the twentieth century, when some irritating young scientists started claiming they had new evidence for the discredited idea that the continents had moved about the Earth's surface.

Chapter 2
The Accidental Geophysicist

Mine was not the most unlikely scientific career by any means, there are plenty of people who have come from less likely backgrounds, but neither was my path laid out for me. I was born to a farming family in rural Australia near the end of World War II. We were about as far from, for example, London and New York as you can get. Our town was not wealthy, many of the locals were what we call 'battlers', and the local school was generally not as good as those in Melbourne, the nearest large city. Many of the kids in my school were rough and tough, not much interested in learning. There was a shortage of teachers as the proliferating baby boomers came along behind me, and some of the teachers who made it to our small country town were not particularly learned either. However there were certainly teachers who cared enough to encourage bright kids along.

Many of the kids left school as soon as they were allowed, at age 14 in those days. From a cohort of about 100 who started high school (Year 7 of our school sequence) only 11 continued through Year 12. Most of those were going on to teachers' college or nursing training. Only two or three of us went on to university and I was probably the only one who had any interest in an academic career.

I was certainly the first in my extended family to get anywhere near university. My father had left school at about age 12, he was needed on his father's farm and he was happy to do that as far as I know. On the other hand my father had a lifelong curiosity about the world and he read a lot. My mother also read a lot, though more fiction than current affairs or history or whatever took my Dad's fancy.

My parents were honest, solid people, but not exactly into sophisticated social graces. Both were from what we call battler's families. Dad's father was a farmer who scratched along. Mum's father was a drover, for most of her growing up I think, so her mother raised five children with only the intermittent presence of their father. In due course I went off to the city and university, smart enough but feeling quite socially awkward.

Books were provided for us and reading and enquiring were encouraged. Astronomy was the glamour science and I became interested. Dinosaurs were a

G. F. Davies, *Stories from the Deep Earth*,
https://doi.org/10.1007/978-3-030-91359-5_2

minor interest, but they were not a major child-entertainment industry in those days, and lot less was known about them.

I was the youngest of three. My sister, seven years older, would, in the nineteen fifties, expect to get married and raise a family. This she happily did, and very successfully. My brother nearly three years older, was good with his hands and became passionate about the farm our father was running. He left school at age 15 to work with my Dad. They were 'progressive' farmers, taking on a lot of the modern methods of mechanical and chemical agriculture on our fairly small farm, and they were good at it. It was much later that the long-term drawbacks of this approach began to emerge.

I was less practical, or less encouraged anyway, and I began to do quite well in high school. The term 'nerd' did not exist in those days, but I was certainly that way inclined. At some stage my interest in astronomy germinated the idea that I would like to *be* an astronomer. This became a general, if distant, goal.

It was, in retrospect, a fortunate time for such an aspiration. The economy boomed through the fifties and the wealth was shared around better than now, so ordinary people's lives were materially improving at a noticeable rate. Education was valued and although the government was conservative for the time it set up a system of Commonwealth Scholarships to assist bright kids to go to university. It also built a string of new public universities to accommodate the baby boomers. I did well enough in our small country school to gain admission to university with a scholarship. I could probably have gone without the scholarship, but it would definitely have strained the family finances for a while, even though fees were not very high in those days. Sadly this system was replaced in the 1980s by politicians who had themselves benefited as I had from government assistance. By now a university degree requires wealth or the accumulation of huge debts, even though our university system is still nominally public.

I could have gone to Melbourne University, a century old, but chose instead the new Monash University. It was only a year old and still mostly open paddocks on the fringes of Melbourne. The learning part was in one corner and my campus residence in the opposite corner, a walk of ten minutes or so. This choice reflected my propensity to avoid traditions that I found stuffy, or that challenged my social awkwardness. I tended to be a bit rebellious, though not enough to get in serious trouble, and I preferred the new and slightly more adventurous.

Monash was on the eastern edge of Melbourne at that time and rather isolated. The nearest suburban train station was a 20–30 min walk away, and it was then a 30-min run into the city centre. There were no food or cultural activities to speak of for miles. There were a few buses but most students drove. It was a commuter campus, this being the early 1960s. Living on campus had the advantage of avoiding commuting and the disadvantage of being an isolated enclave, especially in the early years.

Monash's population virtually doubled every year for several years as the baby boomers piled in—750 my first year, 1500, 3000, 4500 and then a more steady increase. The residence also expanded and another popped up next to it. So the place changed rapidly, which increased the diversity of campus life. Although life was rather different on 'the farm', as Monash was known in its early years, compared with 'the shop' (Melbourne Uni) right next to the city centre, I got quite a lot of the extras that Universities traditionally were supposed to confer, before they became much more banal and expensive degree factories.

I settled into the academic work easily enough, doing the science track of physics, chemistry and two maths, then physics and maths, then just physics in my third year. There was nothing of the North American tradition of 'liberal arts' in which you spent your first two years studying a wide range of humanities and science. Though a country school like mine was not as good a preparation as the better Melbourne schools, public and private, they allowed a bit for that and I wasn't troubled, though it was busy and challenging at times. As we progressed, a pattern emerged in our science cohort that two other students alternated in coming out top while I came in third fairly consistently. I was doing quite well, but was not a brilliant star by any means.

I finished my three-year degree and continued for the 'Honours' year, following the British tradition. In the Honours year of physics the biggest requirement was to do a major essay on some topic of our own choosing, drawing on original research papers. It was a good idea, but more guidance would have helped a lot. I chose a topic I had noticed, but it didn't lead very far. There was really only one paper on it. I struggled to find related papers and construct an essay from limited material. The Honours year was graded simply as I, IIA, IIB, C and don't bother. You needed I (first-class) to get into post-graduate study. You might scrape in with IIA. I got IIA.

I can't say my undergraduate education as a scientist was a great one, in retrospect. I learnt a lot of facts and techniques, but how did one actually 'do' science? One day out hiking, mind coasting, I realised why a funny triangular plot was used in a certain part of chemistry. We had been shown it, but *why* it worked that way was not explained. In physics we were presented with 'the Debye model' of thermal properties of a solid (don't worry, you don't need to know what it is). Years later in my graduate work I discovered why it was useful. In statistical mechanics we were presented with 'the canonical ensemble', then 'the *grand* canonical ensemble' and how to derive them. But how did they come to be? What motivated their invention? What were they for? Not explained.

I had inklings of dissatisfaction with what we were being fed. I think that was the best sign of my awakening as a scientist. I wanted to know why. Over time I learnt to notice when there was a 'why?' coming up, even if I didn't know how to answer it. Over more time I found that answers might eventually show up, if you stayed alert. I didn't do very well in third year physics, after doing very well in the first two years, and it occurs to me only now that they had lost my interest because they weren't motivating what they were feeding us. They were going through the motions they had been taught.

My friend Murray, rival scholar and fellow bushwalker (hiker), said there was a Professor in Applied Mathematics interested in astrophysics and Murray was going to talk to him about doing a Master's degree. That sounded interesting so I said can I come too? Professor van der Borght, not long from South Africa, took us both on. He was a lovely friendly man who loved his mathematics. Murray had a top scholarship to Cambridge so I think he did a short project and left for Cambridge later in the year. For my own Master's project I got into calculating the interaction of rotation with a magnetic field inside a big star.

When I was starting at Monash they had said the physics department specialised in solid-state physics. That didn't sound much like astronomy, but they said that's OK, just do a degree in physics and you can learn the astronomy later. So I inadvertently learnt some physics that was more relevant to what I actually ended up doing. On the other hand my Master's project did introduce me to some fluid dynamics, and that turned out also to be a very good thing.

For my Master's project I carried a deck of computer punch cards, holding the program code, back and forth across campus trying to get my calculation to run on the one 'mainframe' computer on campus. You submitted your box of cards in the afternoon and came back next morning to reclaim them, plus a fan-fold printout of results. You poured over the printout, trying to see why it hadn't worked, revised a few cards and repeated the process. Welcome to the world of computers, very fast and completely stupid. You have to spell everything out *exactly*. It kept doing what I *said*, instead of what I *meant*, which any sensible human would have understood. Eventually it worked and I got some good results that turned into my first published paper, but it was a slog.

It was an interesting experience. I was aware that Prof was an applied mathematician mainly interested in beautiful mathematics. He looked to real-world situations to find interesting problems, but was less interested in how much you learnt about the real world after you had found your beautiful solutions. Most economists have a similar attitude, but with more consequence for humanity. At one point I came to him and said I didn't see how one key part of the calculation could be specified. He thought for a moment and said Oh, OK do it this way then. Later I understood that he hadn't thought the problem through properly, and to do what we set out to do would be much more challenging, or would need information that was not available. No problem, he just switched me to a mathematically easier version that might or might not have a lot to do with a real star.

I suppose it was late that year (1966) I applied to some PhD programs in astronomy around the world, mostly in North America and a couple in the UK, to start in the northern 'Fall', i.e. September the next year. For the hell of it I included an application to Caltech (California Institute of Technology) because they had the biggest telescopes and were about the best. They would accept perhaps five or six students per year and I didn't expect to get in. I didn't. What I did not expect was an offer to do *geophysics* at Caltech. They had a planetary science program bridging between astronomy and Earth science and my folder had been passed around.

I did not know much about what geophysics might be. As it happened my house mate Pete had studied geology in Tasmania and had a volume from 1958 edited by Professor Sam Carey, who was an advocate of continental drift, through the dark ages when that was a very disreputable idea. I had read Carey's paper and found it pretty interesting. Still I should find out more about geophysics.

Geophysics: physics applied to the the Earth. Like astrophysics (stars), or biophysics (living organisms). I found out there was a geophysics group at Flinders

University in Adelaide. I hitch-hiked there to visit them. They were studying ocean waves, the mechanics of creatures swimming, and other things I don't remember. Their field was probably the physical side of oceanography. Meteorology is the study of the atmosphere, mostly physics and some chemistry. Still, the Caltech group seemed to be focussed on the solid Earth, from the crust down. That would include seismology, which involves both earthquakes and the waves they send out through the Earth. It could include the source of the magnetic field, believed to be in the liquid iron core. There might be other topics too. Even though it wasn't what I had imagined since I was a kid, a degree from Caltech in anything was not to be sneezed at. Alas, by the time I felt informed enough to say yes to the offer it was too late. They apologised and said they had offered the place to someone else.

Prof and I contrived to keep me working, finishing my Master's and going on to some potential PhD work with him. So the next year I was still around and applied again to Caltech, this time with a completed Master's thesis. I was again offered a place in geophysics, to start September 1968. I didn't waste any time accepting that offer.

So, by happenstance, I was deflected from my aspiration to be an astronomer. It was probably a good choice for me because astronomy was, and is, a very competitive field and I might or might not have done very well. Geophysics, on the other hand, had just been opened up with a radical new theory and there were lots of opportunities. I was able, over time, to find a niche that suited my strengths. I am not the most brilliant mathematician, though I can do what I need to do, nor do I excel at measurement and observation. Rather, my persistent questioning and a cultivated ability to absorb a lot of information and to find patterns within it were the traits that eventually bore fruit for me.

Many of my friends had gone to the UK, either to study or for the adventure of travelling, or both. (Anne, a keen bushwalker, travelled overland solo, including through the Middle East. It wasn't as murderous then as now, but a cute blonde (feisty) woman travelling solo? They were an adventurous lot, those bushwalkers.) I was different in that I went off to the US, which was probably much better for me. I would not have taken kindly to the stuffy traditions in the UK and to being treated like the colonial, but that issue didn't exist in California.

Chapter 3
A Propitious Time

I duly arrived in Pasadena, part of 'greater Los Angeles', in September 1968, a few days before my twenty-fourth birthday, and found a place in a share house of five graduate students. It was a good group with a lively social life. 1968 was a time of great social upheaval, with hippies, civil rights, women's rights and resistance to the Vietnam war. There was at least one large anti-war demonstration somewhere in the US virtually every week. Before my arrival Martin Luther King and Bobby Kennedy had been assassinated. There were huge demonstrations in many countries demanding change. I learnt recently that President De Gaulle had even briefly fled France fearing an insurrection, as Paris was paralysed by strikes and protests. I recently found some old letters of mine in which I wondered about the wisdom of moving to the US at such a time. I had forgotten it was such an immediate concern. Anyway I went.

It was also a propitious time to enter the field of geophysics. The new theory of plate tectonics had just swept the field and it opened many further questions and opportunities. On the other hand the group I landed in had played little role in the development of plate tectonics. Their main focus was seismology, but with an expanding interest in the properties of materials under high pressures and temperatures. The Seismological Laboratory of Caltech was the most prominent source of information whenever there was an earthquake in the Los Angeles region, which was about every few months. They were practised at dealing with the media. It was where Charles Richter had worked and developed the Richter scale of earthquake magnitudes. He was retired but still occasionally came in to the 'Seismo Lab'.

In the British tradition doing a PhD involved focussing immediately on a research project. On the other hand in the US it was normal to spend a year or two taking more courses relevant to your field before getting seriously into a research project, which might take another two to four years. I had done more maths and physics than most of the local students, who had been through a liberal-arts degree with less undergraduate specialisation. I had also done a research project for my Master's degree. I was keen to get on with some real science. My supervisors compromised and allowed me to do fewer courses but, sensibly, I was required to learn some geology.

G. F. Davies, *Stories from the Deep Earth*,
https://doi.org/10.1007/978-3-030-91359-5_3

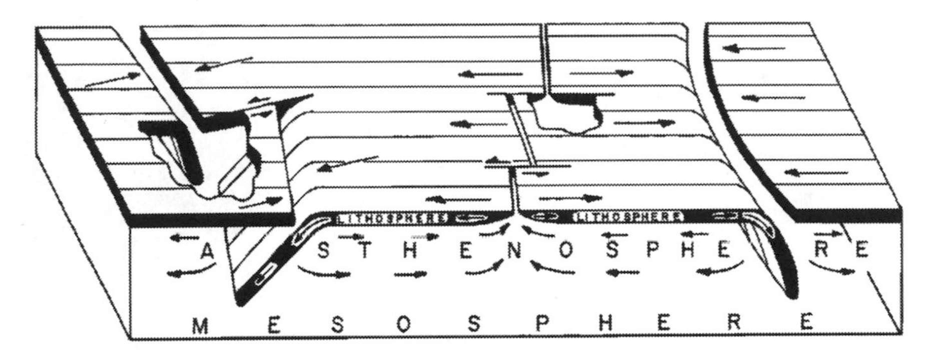

Fig. 3.1 Cartoon of plate tectonics in three dimensions, as envisaged by Isacks, Oliver and Sykes in 1968. Reproduced with permission [2]

It was a good arrangement. As well as doing some course work we were required to do several small research projects before plunging into a major PhD project. I was able to get a couple of those research projects going and I learnt a lot of new stuff. I was exposed to the bewildering array of rock names, each of a particular composition or texture, made of a bewildering array of minerals, each a separate chemical compound but often with a range of composition. The very messy world of petrology, the chemistry and thermodynamics of how rocks crystallise or otherwise come to have the myriad forms they have. Some very basic structural geology, a few fossils and a couple of field trips to see things in their real settings and I was getting acclimatised to a very different subject. That was on top of doing a bit of more advanced maths and learning about wave propagation and some quite sophisticated mathematical methods to deal with it, and so on.

We were introduced to recent papers creating excitement about this new thing called plate tectonics. There were papers by W. Jason Morgan of Princeton University and Xavier Le Pichon of Columbia University showing how plate motions on a sphere can be described by rotations about a point somewhere on the sphere. There was a paper by Dan McKenzie and Robert Parker, of Cambridge, UK, showing that earthquakes around the eastern and northern Pacific rims all coherently reflected motion of the Pacific plate relative to adjacent continents.

Another paper, of prime interest to my new mentors and colleagues, was called *Seismology and the New Global Tectonics*, by Bryan Isacks, Jack Oliver and Lynn Sykes of Columbia University. They showed how many observations, accumulated over many years, were consistent with 'the new global tectonics' and shed more light on it. Their summary illustration is shown here in Fig. 3.1. It shows several moving plates, reminiscent of those in the Pacific ocean region, and their conception of what might lie beneath. They show the plates as pieces of a strong *lithosphere,* about 100 km thick, moving over an *asthenosphere* in which the mantle is soft enough to move slowly, and below that a *mesosphere* presumed to be stiff enough not to be involved in the shallower motions, so it is below the black line at the bottom, about 700 km down.

There is a coloniser effect in scientific ideas, analogous to the coloniser effect in island ecosystems. The first organisms to reach a new island, by whatever chance circumstances bring them there, are free to establish themselves, but later organisms have to compete with those already there. The result is that, even much later, island ecosystems tend to be dominated by the species that arrived first. When a new scientific field is opened up the first ideas proposed about how it works tend to become the benchmark, and later ideas have to compete to establish themselves. This is an example of a more general quirk of the alleged intelligence of human beings, according to Daniel Kahneman in his book *Thinking Fast and Slow*: when we encounter something new we tend to assume our first experience of it is normal, and that any change from that state is an aberration, even though we might by chance have encountered it in an atypical state.

So it was with this diagram. They included features that already had a history and some reasonable evidence behind them, as we will see later. However there was no compelling reason to draw a black line along the bottom and thus confine motions related to the moving plates strictly to the shallower mantle. Nor did the arrows of the 'return flow' under the plates need to be so shallow, they were not based on much fluid-dynamical insight nor any quantitative demonstration. Some other features also were schematic, and would need to be refined. Whereas before there were general ideas, and some debate, about the 'strength' of the mantle, both shallow and deep, once this diagram was published, etched in simple, bold lines, the picture was firmly placed in many minds. It became harder to think and talk in any other way.

On the other hand the elucidation of the plates near the surface, and the large amount of evidence they marshalled, still made the paper a *tour-de-force*. They showed how earthquakes at mid-ocean rises, where plates pull apart, are consistent with that pulling apart. Earthquakes below deep ocean trenches are consistent with plates pushing together and one of them plunging back into the mantle. They even showed that seismic waves at a *subduction zone*, where one plate is going down, defined a zone of more efficient propagation of seismic waves which they inferred to be the descending lithospheric 'slab' itself, cooler and stiffer than surrounding mantle. Above and outside the slab was a zone of less efficient propagation, inferred therefore to be hotter and the source of volcanism that is a feature of many subduction zones. These features had already been discerned under the Japanese islands by Japanese seismologists, notably T. Utsu in 1966.

The idea of seafloor spreading, where plates pull apart, had been proposed earlier in the decade, by Robert Dietz in 1961 and Harry Hess, whose paper was published in 1962 but had been circulating for a couple of years previously. Hess and Dietz, it is vouched by others, reached their ideas independently.

The idea of subduction was less explicitly developed before this time, but there were precedents. K. Wadati in 1931 documented the presence of very deep earthquakes under Japan, defining a zone starting at the Japan trench and sloping down to the west and reaching a depth around 600 km. Hugo Benioff at Caltech later documented deep seismic zones in other places. For a time these became known as Benioff zones, but later the role of Wadati was recognised by calling them Wadati-Benioff zones.

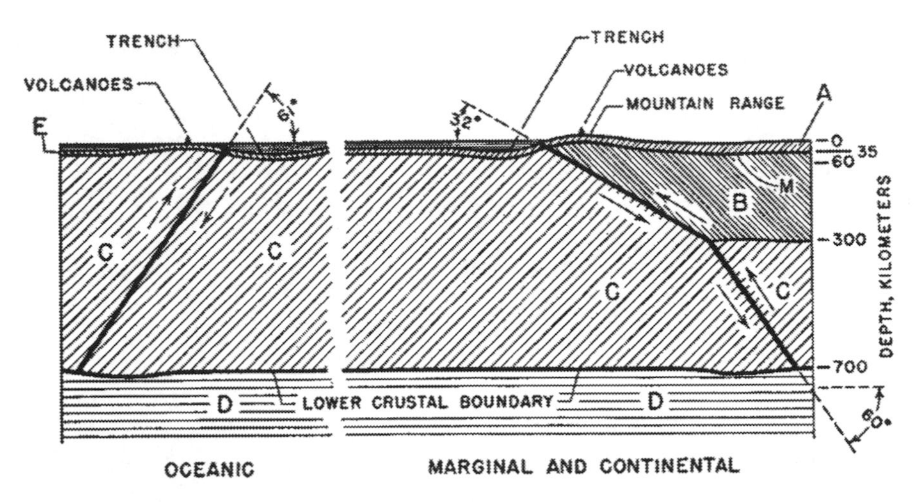

Fig. 3.2 Benioff's 1955 interpretation of Wadati-Benioff zones. The structures seemed to differ between purely oceanic regions like the Mariana Islands of the western Pacific (left) and the edges of continents (right). He assumed a 'crust' that was strong and brittle down to 700 km (regions C). Mantle deeper than that was taken to be fluid or plastic enough to prevent the accumulation of stresses that would cause an earthquake. Earlier evidence had already indicated a deformable 'asthenosphere' starting only 100 km down, as depicted in Fig. 3.1. Fair use: from Benioff 1955 [3]

In the 1950s Benioff had interpreted these zones as faults cutting deeply into the mantle, which he assumed to be solid to a depth of about 700 km. He inferred that the faults reflect horizontal compression across the faults. A summary diagram from a 1955 paper is shown in Fig. 3.2. Benioff interpreted the deep ocean trenches, typical of where the zones reach the surface, as resulting from the under-thrusting of the oceanic side, but seemed to conceive of only a limited amount of displacement on the fault, by perhaps kilometres or at most tens of kilometres. This interpretation was also supported by some remarkable measurements of the gravity field over the Indonesian trench by F. A. Vening Meinesz in the 1930s using submarines: the slightly weaker gravity implies a mass deficiency due to downwarping of the surface.

A bolder interpretation by American geologist Reginald Daly, in a 1940 book called *Strength and Structure of the Earth*, apparently went unnoticed. Daly thought compression of the Wadati-Benioff zones would force a 'prong' of lithosphere to project down into the asthenosphere, which was established as a plausible concept early in the twentieth century. The work summarised by Isacks, Oliver and Sykes (Fig. 3.1) gave new support to the idea. An even bolder picture, involving mantle convection, had been proposed by Arthur Holmes in 1929. We'll be looking at those ideas in a later chapter.

The proposal of seafloor spreading by Hess and Dietz in the early 1960s led to a recognition that a complementary removal of surface material would avoid the implication that the Earth had to be rapidly expanding, an idea that had been advocated by Sam Carey among others. Although Hess did not develop this part of his picture as much, the implication was clearly made that seafloor 'shrinking'

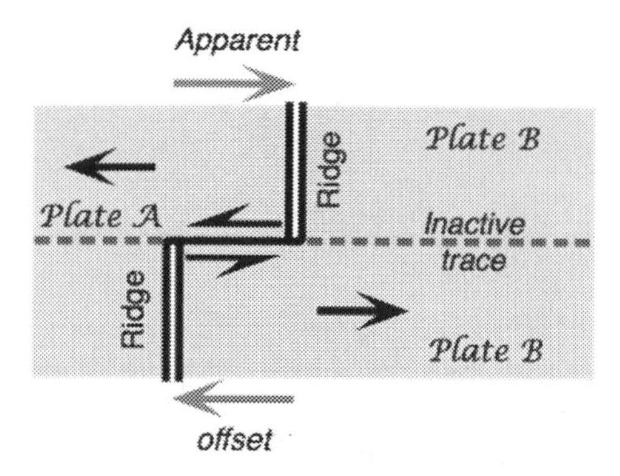

Fig. 3.3 Sketch of a left-lateral transform fault joining two ridge segments. Two plates (*A* on the left, *B* on the right) are pulling apart at the ridge segments. This means the 'northern' side of the transform fault is *Plate A*, moving to the left, and the 'southern' side is *Plate B*, moving to the right. Thus the motion across the fault is left lateral, in geologists' classification, which is opposite to the apparent offset of the ridge segments. Beyond the ridge segments on the right, both sides of the trace of the fault are *Plate B*, and there is no relative motion, so this is just an inactive trace or scar left by the active fault segment

occurred at a trench to balance seafloor spreading at a mid-ocean rise. This is made explicit in Fig. 3.1 above.

A third kind of boundary between plates occurs where two plates slide horizontally past each other along a 'transform' fault. The papers of 1967–68 attribute the concept of a transform fault to a 1965 paper by J. Tuzo Wilson. A key realisation of Wilson's was that the relative motion where one of these faults offsets the crest of a mid-ocean rise is opposite to what it at first appears. In the example in Fig. 3.1 the 'spreading centre' (or mid-ocean rise or ridge) is offset to the right as you move along the rise crest. Figure 3.3 shows the situation in more detail. If you look closely at the relative motion of the plates across the fault you will see the opposing plate is moving to the left—the fault is 'left-lateral' as geologists say. Furthermore the relative sliding is confined to that part of the fault lying between the rise crests, and beyond that segment the fault would be merely an inactive scar.

Wilson had pointed out that maps of earthquake locations in the oceans showed them indeed confined to the fault segments between crests, and he took this as strong evidence in favour of his interpretation. Lynn Sykes, in work summarised in the paper by Isacks, Oliver and Sykes, had confirmed this finding and added to it the fact that waves from these earthquakes could be analysed to confirm the sense of slip, left-lateral in the case illustrated. Wilson's interpretation of transform faults provided strong and immediate evidence for the reality of seafloor spreading.

I don't recall when I first read Wilson's 1965 paper. It would have been some years later, perhaps many years. What I do know is that there came a time when I read it carefully and realised it is much more than a paper about a new kind of fault, important as that is. Its full title is *A new class of faults and their bearing on continental drift*. Although much of the short paper is devoted to explaining transform faults, it is also the first time the complete concept of plate tectonics was presented to the world in all its simplicity. The logic was clear, concise and compelling. He even coined the term *plate*, and presented a sketch map of the global plate system, bounded by a fully connected network of rises, trenches and transform faults.

The papers I was first introduced to at Caltech had, it seemed to me, failed to perceive or fully acknowledge Wilson's historic contribution. This was the paper that finally made sense of many puzzling observations of the Earth that had been accumulating for several centuries.

Before going into that history, I will digress back into my own experience of my early years at Caltech. One of my small research projects, which we were required to do before we launched into a main PhD project, was to see if the record of large earthquakes since about 1900 was consistent with the predictions of plate tectonics. Plate directions and velocities had been estimated from the record of magnetic anomalies around mid-ocean rises, which we'll look at in a later chapter. According to my supervisor, seismologist Jim Brune, it should be possible to use the record of earthquake magnitudes to calculate plate velocities at oceanic trenches, where one plate was supposed to be plunging below another, back into the deep mantle.

The biggest earthquakes resulted from slip by several metres of one plate past the other over fault areas extending for a few hundred kilometres. With some simple formulas and estimates we could convert earthquake magnitude into an amount of fault slip. This I did, and got results that were roughly in accord with predictions—a few centimetres per year, more for the more active zones like Japan. In due course we wrote a paper which became my second-ever Earth science paper and third ever paper. It received quite a lot of attention and remains among my more cited papers.

A couple of things are worth mentioning about that paper. It soon transpired that the formulas we used are not very good for the largest earthquakes, which account for most of the fault slip. The problem is that you need to look at seismic waves with wavelengths at least as big as the earthquake rupture, in order to capture the total amount of slip. The older magnitude scale we used was based on waves of only around 150 km wavelength, but the very largest 'great' earthquakes break the fault for many hundreds of kilometres. So our numbers under-estimated the amount of slip. Nevertheless they showed a rough consistency with plate tectonics and helped to establish its plausibility.

With better seismometers capable of recording those very long waves, a new magnitude scale was defined. Then the biggest earthquakes were recognised to be over magnitude 9. The biggest recorded are Chile, 1960, 9.5; Alaska, 1964, 9.2; Sumatra, 2004, 9.1; Japan, 2011, 9.0; Kamchatka, 1952, 9.0.

Twenty five years later, reading up on the history of the subject for the book about mantle convection I was working on by then, I came across the metaphorical footprints of Charles Darwin. During his great voyage in the 1830s he experienced an earthquake in Chile that he learnt had generated a fault scarp of 1–2 m displacement. Darwin calculated that with an earthquake like that happening only about once a century (near the actual rate) then the Andes mountains could have been raised in rather less than a million years. This was an influential estimate, because it helped to establish the plausibility of Charles Lyell's hypothesis that you could explain geological formations just by the long-term action of presently observable processes.

It is my experience, and an observation of others, that many modern scientists have only a limited knowledge of the history of their subject. My ignorance of Darwin's calculation over a century earlier is not particularly remarkable, as Darwin's point was important at the time but soon overtaken by an abundance of other evidence. Besides I was new to the subject, although my supervisor would have been expected to point out relevant important precedents. In fact he did point out to me the papers by Benioff a decade or two earlier on what by then we were calling subduction zones; Benioff had attempted a version of what we did, adding up the cumulative deformation in an earthquake zone, but the measure we were using was better-based and had a simpler interpretation as an amount of fault slip.

The under-stating of Tuzo Wilson's 1965 paper by authors writing only a couple of years later is perhaps also understandable, though still a little troubling, because many of the key ideas were already very current among those doing immediately relevant work. For example, Lynn Sykes was explicitly testing the idea of seafloor spreading using earthquake locations and mechanisms. For them, Wilson's idea was also the key to a more integrated understanding. On the other hand Wilson was not directly connected with the groups most actively involved in observational work, and they may not have been aware of his other recent papers building a comprehensive case for continental drift, and adding the seminal idea of what Wilson called mantle hotspots. So I suspect Wilson's contribution was not fully comprehended by many busy, competitive scientists rushing on to their next publication and their next funding application.

A more telling example is that by about 1945 most of the key ideas were available that could have led to plate tectonics, or at least to a global scheme of mobile crust, but hardly anyone came close, as I noted in my 1999 book *Dynamic Earth*. The closest approach was by Arthur Holmes in 1928, when he proposed that continents might be carried along by convection in the mantle below. He included the idea in his book *Principles of Physical Geology*, first published in 1944 and widely-used in the UK. Reginald Daly, in his own book mentioned earlier, got something close to subduction. Had Darwin's calculation been better known, perhaps Benioff would have been bolder and looked at the possibility of repeated thrusting earthquakes over a geologically substantial period. That would have implied that sea floor was consumed at deep ocean trenches.

Another factor was at work through the early and middle parts of the twentieth century, and that was a certain timidity. It seems to me that people had stopped asking the big questions, and preferred to stay close to their own specialty. Compounding this was that Alfred Wegener's idea of continental drift, first formulated in 1912, had become very disreputable, at least in the anglophone world.

I have presumed this timidity to be a product of the increasing specialisation of the subject, but Marcia Bjornerud suggests it may also owe something to a big conflict between geology and physics in the latter half of the nineteenth century, that we'll come to in the next chapter. Some geologists developed a hostility to physics that did not serve the subject well. Wegener was both a physicist and a German, so his ideas were doubly anathema in the Anglophone world.

The effect of this timidity can even be seen in Fig. 3.1 and the discussion of it in the original paper. The authors considered several reasons why lithosphere, as traced by deep earthquakes, descended only several hundred kilometres into the mantle. One possibility they entertained was that plate tectonics had only been operating for as little as ten million years. Alternatively it was suggested plate tectonics had operated for longer, but not earlier than the Mesozoic (about 250 million years ago). Another possibility was that a plate might stop moving when its leading edge encountered the mesosphere. A third possibility was that the 'slab' was absorbed back into the mantle to the point of not sustaining earthquakes, implying a re-absorption time of a few million years; it would then be a coincidence that in a few subduction zones the deepest earthquakes reached the depth in the mantle where its properties were known to change due to the increasing pressure. In any case in each of their diagrams the descending slab terminated rather abruptly, following the rather abrupt cessation of deep earthquakes.

The authors did not seem to be willing to contemplate that plate tectonics might have been operating for, say, billions of years and that a great deal of lithosphere would thereby have been cycled back into the mantle. That would make the mantle a product of a long-standing dynamical process. Perhaps it is unfair to those particular authors, who allow they were very much getting used to this new perspective on the Earth's workings. Yet even a decade later few seemed to be willing to contemplate the implications of such an extended history. There also seemed to be a tendency, in this and many later papers by others, to concede only a minimum depth of the mantle to be actively involved with the plates. The old rigid mantle was abandoned only by degrees.

As it happened I was not to continue exploring these topics for some time yet. My project on slip rates in subduction zones was completed and I moved into a rather different topic for my PhD, namely the properties of materials under the very high pressures and temperatures of the deeper mantle. A primary reason, it is pertinent to record, is that the relevant professor, Tom Ahrens, had lots of grant money (from military sources) and could afford to support several graduate students.

This was not a problem for me. I was keen to get on with research and they were happy for me to do so. I gradually focussed in to a rather narrow topic, the appropriate way to formulate the descriptions of large deformations in a way that was 'frame-indifferent', in other words that did not depend on the coordinates or frame of reference used. It sounds arcane, but it was needed to properly describe the elastic properties of materials that have been compressed by large amounts, and the topic had been confused. Probably most PhD topics are very narrow—you are an apprentice, and you approach the frontiers of knowledge at first on a very small front. Later you might broaden out, to a greater or lesser extent.

The Seismo Lab of Caltech was a stimulating place. There were regular coffee breaks to which many of the troops came, and there were often animated discussions of recent work or new papers. If there was no such discussion the Director, Don Anderson, would make a provocative statement to get one going. I learnt a lot and enjoyed the repartee. My fellow graduate students were very much a part of the stimulating atmosphere. At one point some students claimed that we had all the good ideas and the Professors were only there to raise the research grant money. The Professors smiled and did not bother to dispute the claim. It was, after all, their job to turn us into independent scientists, just as a parent's job is to cultivate an independent new person. Not all graduate programs were as encouraging, but perhaps we were a bit too cheeky.

Chapter 4
Water, Heat, Time, Mountains

The modern science of geology is regarded as having emerged in Europe in the eighteenth and nineteenth centuries, though with earlier precursors. I want here only to touch on some highlights of that development, so as to emphasise that the ideas of plate tectonics and mantle convection were built on a great deal of work by many people. They also illustrate that the authors of key ideas, including ideas later discarded, were generally very smart people grappling with puzzles as best they knew how.

It is too easy to dismiss as ridiculous fantasy the idea, for example, of a great ocean that gradually subsided as all the rocks crystallised from it, but sensible evidence was brought to bear. That evidence was, from our perspective, very sketchy and incomplete, but our evidence about the workings of the Earth at remote depths and times is always incomplete. Nor was everything claimed for this theory later discarded. People were asking big questions, and looking for answers in terms of a world governed by patterns or 'laws' rather than by the supernatural acts of gods. Anthony Hallam's *Great Geological Controversies* (Oxford, 1989) is a readable source for this discussion.

A key step was taken by Abraham Werner, who attained a position at the Freiberg Mining Academy in 1775. Werner had grown up in a mining district and set about classifying the many minerals and rocks known at the time, and he is reputed to have made substantial contributions in that way. However he went much further and proposed a comprehensive theory of how all the rocks and geological formations came to be. He was widely reputed to be an inspiring teacher, so although he did not write a great deal his ideas were spread by many enthusiastic students.

Werner drew on previous work that had identified various kinds of geological formations and strata. He proposed a universal sequence, and proposed that they were laid down by a great, turbulent ocean that initially covered all the land but whose level gradually lowered, depositing the various kinds of rocks as it subsided. The first, *primitive* rocks were formed by chemical precipitation, having been dissolved in the ocean waters. These include granite, gneiss, schist and others. There were then *transitional* strata that included limestone and diabase, which we now identify as

G. F. Davies, *Stories from the Deep Earth*,
https://doi.org/10.1007/978-3-030-91359-5_4

a shallow intrusive equivalent of basalt. Then there were stratified groups and less consolidated groups supposed to be increasingly the result of mechanical deposition from suspension, rather than chemical precipitates. Finally there were volcanics, including lavas and tuffs.

Superpositions of these groups were evident in Werner's native Saxony and other places in the region, so there was an observational basis. It was supposed that the primitive rocks formed the cores of mountain ranges, but their upper surfaces were very uneven due to erosion by the early turbulent ocean. It was known that some strata are inclined to the horizontal, some even nearly vertical, and this was explained as due to them having been crystallised onto inclined surfaces. Curiously Werner distinguished *basalt*, occurring as flat-lying strata in Saxony, from *lava*, known elsewhere to emerge from volcanos. There was a long debate concerning whether or not they were similar or identical rocks. The debate came to be known as *neptunists versus vulcanists.*

Volcanos clearly required heat to have melted rocks. Lacking any other source of heat in his grand scheme, Werner proposed that volcanos were a recent phenomenon that resulted from combustion of underlying coal seams. He had witnessed the consequence of a coal-seam fire elsewhere in Germany the previous century, including rocks baked by the local heat.

Gradually more people concluded that the basalt and lava were the same thing. Another key finding was of inclined conglomerate strata in the Alps comprising a mixture of sand and pebbles. It was impossible to conceive of the pebbles having been stuck to the inclined slope by natural processes, and many were persuaded that the strata must have been laid down horizontally and later tilted by forces unknown. This shed a different light on highly contorted strata within the Alps, suggesting that the contortions resulted from later forces rather than being part of their original formation in the great turbulent ocean. A further difficulty was that Werner had no explanation for where all the water went from his great ocean.

Meanwhile in Britain James Hutton, having been a farmer and successful small businessman, acquired the means to move to Edinburgh to pursue scholarly interests. Hutton argued that both basalt and granite were formed from a melted or magmatic state, whereas Werner had claimed both to be chemical precipitates from solution. Hutton devoted some energy to demonstrating that granites were not deposited from above but intruded from below. One kind of evidence he found was dykes of granite leading up into a larger granite body. A dyke is a near-vertical sheet of rock cutting across pre-existing rocks, typically because magma forces open a crack into which it then flows.

However this was not a point that Werner's followers conceded easily, since Werner also allowed for dykes, formed in cracks within pre-existing rocks. The question was whether the cracks were filled by material falling from above or by magma injected from below. The presence of baked rocks in contact with the dykes was one important line of evidence for the involvement of heat rather than water. Because of the typically large volumes of granitic bodies and the implication of great heat at depth in the Earth, Hutton and his followers became known as *plutonists.*

By the accounts of his friend John Playfair, Hutton was a highly animated person and somewhat charismatic like Werner. However he seemed to be incapable of setting his ideas down coherently. His 1795 *Theory of the Earth* is reported to be verbose and difficult, and Playfair reports that he even left out some of the critical field evidence for his theories (though there is a suggestion that illness might have interfered with his writing). After Hutton's death in 1797, Playfair in 1802 published *Illustrations of the Huttonian Theory* which is reported to be far more articulate and comprehensible.

So Hutton vigorously challenged Werner's scheme and a debate raged among their followers. Evidently Hutton went to the other extreme of attributing all consolidation of rocks to the action of heat, whereas Werner appealed almost totally to the action of water. It is not uncommon for innovators to carry their ideas too far. Today we would say that both are important. Igneous and metamorphic rocks require the action of heat, whereas 'low-temperature' sedimentary formations are mostly due to mechanical deposition from suspension in water. Furthermore many important ore bodies have been precipitated from aqueous solution, though from high-temperature, high-pressure water deep in the crust.

Hutton is in fact most famous among geologists for two radical ideas, often conflated. He proposed that sedimentary formations could be explained by the action of presently observable processes acting slowly over immensely long time spans. He further proposed that the observable formations reflected what we now call a steady-state process. The appeal to observable processes, called *actualism* in continental Europe, was not original to Hutton, but according to Hallam his use of it to infer time periods was original. He was even bolder in conceiving of repeated cycles of deposition, consolidation, uplift, tilting, erosion and more deposition. The cycles revealed 'no vestige of a beginning, no prospect of an end', in Hutton's most famous words, from a paper in 1788.

It was in that year that Playfair accompanied Hutton to the coastal location of Siccar Point to observe a now-famous *unconformity,* where sub-horizontal Devonian sandstones rest on near-vertical Silurian slates (which Playfair called schistus). He wrote:

> We felt ourselves necessarily carried back to the time when the schistus on which we stood was yet at the bottom of the sea, and when the sandstone before us was only beginning to be deposited, in the shape of sand and mud, from the waters of a superincumbent ocean. An epocha still more remote presented itself, when even the most ancient of these rocks, instead of standing upright in vertical beds, lay in horizontal planes at the bottom of the sea, and was not yet disturbed by that immeasurable force which has burst asunder the solid pavement of the globe. Revolutions still more remote appeared in the distance of this extraordinary perspective. The mind seemed to grow giddy by looking so far into the abyss of time ...

Evidently Hutton was already conceiving of this juxtaposition of two sedimentary sequences as being only part of a longer series, examples of 'revolutions still more remote'. Playfair and Hutton did not have a clear quantitative measure of the time intervals they were contemplating, but they knew they were dealing with periods vastly greater than the thousands of years commonly believed at the time.

It was Charles Lyell, working in the first half of the nineteenth century, whose work provided a basis for quantitative estimates of the elapse of time recorded in the crust, though Lyell himself evidently did not make such estimates. Lyell is famous for expounding and applying systematically the idea that geological structures might be explained solely by the slow action of presently observable processes, though the idea was not original to him, as already noted. Lyell published three volumes of his *Principles of Geology*, starting in 1830. It was quickly recognised as an important work, and became regarded by some as the founding work of modern geology. That, we have already seen, is an overstatement, but it was highly influential.

Lyell, and many others subsequently, made use of observations that could be related to historical records, such as erosion rates and deposition rates, and of stratigraphic relationships, to demonstrate that a great expanse of time was required. An eloquent example comes from an address by Lyell in 1850 (quoted by Hallam) even though it is still rather qualitative.

> The imagination may well recoil from the vain effort of conceiving a succession of years sufficiently vast to allow of the accomplishment of contortions and inversions of stratified masses like those of the higher Alps; but its powers are equally incapable of comprehending the time required for grinding down the pebbles of a conglomerate 8000 feet [2650 meters] in thickness. In this case, however, there is no mode of evading the obvious conclusion, since every pebble tells its own tale. Stupendous as is the aggregate result; there is no escape from the necessity of assuming a lapse of time sufficiently enormous to allow of so tedious an operation.

'... every pebble tells its own tale' is such a poetic and telling phrase.

Charles Darwin made one of the first quantitative estimates of the lapse of geological time, in the first edition of his *Origin of Species* in 1859. This was an estimate for the time to erode a particular formation in England, and Darwin's estimate, not intended to be anything more than an illustration, was 300 million years. Though it might have been only rough, Darwin's estimate conveys the idea that the time spans involved in geology, that can be characterised qualitatively only by vague terms such as 'vast', are not 300,000 years and not 300 billion years, for example.

That last sentence may seem to be a bit frivolous, but it expresses an aspect of science whose importance dawned on me only slowly in the course of my own career. Scientific measurements are usually described in terms of their accuracy, but it is not the number itself (in this case 300 million years) but the bounds of uncertainty or plausibility that really convey the important information. Let us suppose Darwin's estimate was accurate to within a factor of three (which it plausibly was). In other words the time lapse might have been as short as 100 million years or as long as 900 million years. That may seem like a great deal of uncertainty, but what did anyone know before such estimates were made? For all anyone knew the time might have been 300,000 years or 300 billion years, in other words *a factor of one thousand* larger or smaller, not just a factor of three. The real work of an estimate like this is to narrow the bounds of our ignorance.

To elaborate the point, one can ask 'what is the shortest plausible lapse of time' and 'what is the longest plausible lapse of time'. Before Darwin's estimate, and others like it, there were no answers to these questions. After the estimate (and my supposition of its uncertainty) there were definite answers (though still with the qualifier 'plausible'). A laboratory physicist or chemist used to dealing with uncertainties of parts per billion might be unimpressed by the actual numbers, but that is not the point. The point is we have learnt something definite.

Let me cite two other examples. In the course of my own work I encountered the idea that there had been one or more 'magma oceans' during the formation of the Earth, caused by the impact of large bodies onto the growing planet. How long would it take for a magma ocean to cool and crystallise? I could see a tendency among others to assume they were long-lived. Eventually I encountered a formula that allowed for a simple estimate. The estimate was a few thousand years. As it is believed the Earth took millions of years to grow, perhaps 30 million years, this result implied that a magma ocean was a quite transient feature of the growing Earth. Much of the time the planet's surface would have been solid. So a quite rough estimate, that might have been wrong even by a factor of ten, yielded an important insight. When you are venturing into fields of ignorance even a little bit of knowledge can be very important.

A second example comes closer to home. Climate scientists have issued reports on the expected course of global warming this century, and the amount of sea level rise we might expect by 2100. Earlier estimates tended to be in the range of 0.2–0.6 m. But James Hansen asked a different question. What is the highest plausible rise? He noted observations suggesting sea level rise was doubling every decade or so, in other words increasing exponentially, and geological records from the previous inter-glacial warm period. He suggested the rise could be many metres by 2100. He was castigated for being alarmist, even by other scientists, but no-one convincingly refuted his logic and his estimate. Hansen has written about what he calls 'scientific reticence', a fear of venturing too far from the conventional and endangering one's professional reputation among one's peers. The question he posed was legitimate. The scientists writing the official reports asked 'what is our most plausible estimate based on well-established science?' Hansen asked 'what is the largest plausible rise?' It is the answer to Hansen's question that is critical to our civilisation.

I have digressed, so let us return to the question of the durations of time involved with geological processes. The debate about this is known as *catastrophists versus uniformitarians*, but it was not so simple as pitting Biblical literalists against radical scientists. Many prominent people of the time professed *deism*, as distinct from *theism*. The deist view was that there was a Creator, but after the world was set in motion the Creator no longer intervened, content to let the world unfold according to laws laid down at the beginning. This left room for natural philosophers to study the natural world, endeavouring to discern God's laws.

Hutton, and Lyell after him, inferred great amounts of time from observing sedimentary strata, combined, evidently with some knowledge of slow rates of erosion and deposition in historical time. In the passage from Lyell quoted above it is clear that Lyell also envisaged mountain building and its associated uplift and contortions of rock formations as involving great spans of time.

It is notable though that Hutton and Playfair apparently had not yet conceived of mountain building as a slow process. This is evident in the language of Playfair's account of Siccar Point, referring to a time when the most ancient stratum 'was not yet disturbed by that immeasurable force which has burst asunder the solid pavement of the globe'. Hutton, in his 1788 paper, refers to '… that enormous force of which regular strata have been broken and displaced; … strata, which have been formed in a regular manner at the bottom of the sea, have been violently bent, broken and removed from their original place and situation.' The language in both cases suggests sudden, even catastrophic change.

Strata in the high Alps were known to be highly contorted, even inverted, and it was difficult to conceive of how such distortions could result from any known contemporary process. Thus it was a common view that there had been one or more earlier periods of Earth history in which far more rapid and violent processes had operated.

Such a view was reinforced as the sedimentary and fossil records became better explored, and used to establish correlations from region to region. Georges Cuvier, of Swiss origin, worked in the Paris Basin (the geological name refers to a body of sediment reflecting a long period of subsidence, rather than to a presently-existing depression). He and colleagues were able to demonstrate a series of cycles of alternating fresh-water and salt-water deposition, indicating a repeated cycle of uplift and subsidence. A striking feature of the marine fossils in these sequences (and in the whole fossil record since documented) is that each marine episode includes a distinctive assemblage of fossil creatures rather different from the previous episode. A strong implication seemed to be of long periods of quiescence or slow change separated by sudden and catastrophic change.

Lyell, on the other hand, argued that all geological observations could be explained by the prolonged action of presently observable processes. A principal influence on the development of his ideas were his observations in Sicily of progressive accumulation of volcanic deposits around Mt. Etna and of associated progressive uplifts of strata containing marine fossils. He conceived, if only vaguely, that volcanism and associated earthquakes provided a mechanism that could produce large uplifts and distortions of strata. Thus he was able to argue not only that sedimentary deposition was protracted but that 'disturbance' was also protracted.

Lyell had also observed a formation in the Paris Basin that comprised many fine bands and contained freshwater fossils of extant shells that he recognised. He inferred that each band reflected a year's deposition, and thus inferred that the whole formation reflected several hundred thousand years of deposition in calm, freshwater lake conditions. Not only were conditions calm but the water was always shallow, implying that the land was slowly subsiding as the sediments accumulated.

It was evidently easier to assemble observations and arguments for prolonged deposition than for prolonged deformation. A major reason for this is that the mechanism of erosion and deposition is more accessible to common observation than is the mechanism of mountain building. Thus in the passage quoted above Lyell appeals to deposition to defend his notion of prolonged process, rather than to the 'contortions and inversions' of disturbance, because 'every pebble tells its own tale'; and the tale told by the pebble is intelligible to the thoughtful lay observer. This passage is from 1850, twenty years after the first publication of Lyell's *Principles of Geology*.

The difficulty of explaining deformation is witnessed by the fact that even a century later conventional geology still had not produced a persuasive mechanism of mountain building. There were theories, but they contended with each other, each had gaps or weaknesses, and none had carried the day.

This is not to say that prolonged deformation has only recently been established. It has been established by documenting the prolonged deposition of shallow sediments adjacent to mountain ranges, similar to Lyell's example just mentioned. The sediments did not progress from deep deposition to shallow, as they would have if there were an initially deep depression. The implication is that the land was slowly subsiding and sedimentary deposition kept pace with the subsidence so that the water was always shallow. Sediments could also be recognised as coming from adjacent highlands, implying that the highlands were being slowly uplifted as the depression subsided. The general point is that geologists could not as easily appeal directly to a mechanism for deformation as they could to the mechanism for grinding pebbles.

Charles Darwin was a fan of Lyell's and took Lyell's treatise with him on his great voyage on the Beagle. He provided several early examples of both indirect and direct arguments for slow change. While in South America he observed a correlation between the elevation of strata in the Andes mountains and the declining proportion of extant species in the fossils contained in the strata. From this he inferred that the higher strata were older, implying the progressive uplift of the strata and thus, indirectly, the progressive uplift of the mountains.

Darwin's experience of an earthquake has already been mentioned, and it complemented his observations of elevated marine fossils. He inferred that such an earthquake, recurring about once per century, would be sufficient to raise a mountain range like the Andes in much less than a million years. His estimate of the time to erode a British formation has already been mentioned.

Darwin's arguments were influential, since they supported an alternative to the catastrophist presumption of the time that the Andes had been raised in one great convulsion. In retrospect they are very significant also because they foreshadowed a profound insight that could have come from the study of earthquakes, but that came instead from another branch of geophysics. We will come to that later.

There were difficulties in Lyell's account of geology. The large change in fossils between the Cretaceous and the Tertiary he supposed was due to the absence of a long segment of the rock record, either because it was never deposited or because it was subsequently eroded away. This is a common concern in geology, and numerous examples could be cited where significant sections of the record are missing. Lyell supposed the missing segment would show a gradual change in the fossils instead of the abrupt break. The absence of mammals in older strata Lyell similarly attributed to the supposed erratic nature of deposition and preservation, such that the older rocks containing mammals had not been found or mammals were not everywhere present on the planet at all times, depending for example on variations in climate. These can be seen as arbitrary assumptions that might or might not be true, more evidence would be needed to decide.

As is frequently true of great innovators, Lyell went too far. He argued that catastrophes played no part in forming the geological record and, following Hutton, that the Earth was in a steady state. He claimed that the record revealed no discernible change in rates of process, or kinds of process, or even in kinds of organisms. He maintained this position in the face of rapidly accumulating evidence from the fossil record that there had been dramatic progressive changes in life forms. This is not to diminish his great achievements, but to illustrate that he was human like the rest of us, and in the heat of the debates he overstated his case. Not everything he said was gospel.

Unfortunately, Lyell was so influential that his work has been treated almost as gospel by some geologists. His 'principle of uniformitarianism' was elevated almost to the level of natural law by some, and arcane debates could be found on precisely at what level of method, process or 'law' his principle applies. It is really rather simpler than that. He was arguing against the natural and prevailing assumption of his time that dramatic results necessarily required dramatic causes.

Catastrophes have become more respectable in recent decades. At the lower end of the scale, stratigraphers have recognised that many sedimentary deposits are biased towards unusually large and infrequent events like exceptionally large floods, which tend to carry a disproportionate amount of sediment. At the medium scale, there was a series of catastrophic floods at the end of the ice age in the western United States, well documented by J. H. Bretz in 1969. Those floods occurred because melt water from the great North American ice sheet was episodically dammed by ice barriers that occasionally gave way, releasing great torrents across the land. There was great resistance to this interpretation of the dramatic landforms left by the floods, such as gigantic 'ripple' marks, because it violated Lyell's so-called principle.

At the largest, planetary scale it has by now become widely accepted that the extinction of the dinosaurs at the end of the Cretaceous period, 65 million years ago, was caused by the impact of a large meteorite into the earth, possibly aggravated by a series of giant eruptions of *flood basalts* in India that was occurring at the time. Other 'mass extinctions' are now associated with other meteorite impacts or flood basalt eruptions. Thus Cuvier's inference of a series of catastrophes separating long periods with distinctive fossil assemblages is by now generally supported, though it took a century of debate and dramatic new evidence to swing the balance.

On the other hand there has long been strong evidence that much mountain building was a slow, incremental process, much as envisaged by Lyell and Darwin. Furthermore we finally have a mechanism capable of driving such slow disturbance. Thus we find that both sides of the debate between catastrophists and uniformitarians had some truth on their side, as we have already seen for neptunists *versus* plutonists.

William Whewell, quoted by Hallam, stated the issue very clearly in 1837, not long after Lyell's first volume was published.

> *Time*, inexhaustible and ever accumulating his efficacy, can undoubtedly do much for the theorist in geology; but *Force*, whose limits we cannot measure and whose nature we cannot fathom, is also a power never to be slighted: and to call in one, to protect us from the other, is equally presumptuous to whichever of the two our superstition leans.

If you wanted to choose between time and force, you needed more evidence than was available at the time. Whewell also addressed Lyell's uniformitarian principle. Pointing out that geological events come in a great range of magnitudes and recur with a great range of frequencies, he notes elsewhere.

> In order to enable ourselves to represent geological causes as operating with uniform energy through all time, we must measure our time in long cycles, in which repose and violence alternate; how long must we extend this cycle of change, the repetition of which we express by the word *uniformity*?
>
> And why must we suppose that all our experience, geological as well as historical, includes more than one such cycle? Why must we insist upon it, that man has been long enough an observer to obtain the *average* of forces which are changing through immeasurable time?

We do now have evidence of long-term change in the geological record, even including a 'vestige of a beginning'. Current understanding of the Earth is that it was indeed internally hotter and more active after its formation, and it has been cooling and slowing ever since, though at a much slower rate than was conceived back then. In the terms posed by Whewell, we have not perceived even one cycle of a steady state process, and are not likely to.

On the other hand Hutton's hypothesis of a steady state is quite a good approximation for the Phanerozoic era, the period from 540 million years ago in which fossils of macroscopic, multicellular organisms are preserved, and which comprise that part of the rock record they were able to interpret in the nineteenth century. We see in the sedimentary record of this era repeated periods of slow deposition and consolidation alternating with episodes of uplift and deformation, mostly proceeding slowly by current understanding. The process was more episodic than cyclic, and it was disrupted five times by major reorganisations of life, but it then resumed more or less as before.

When I first read of the early debates in geology I was surprised that anyone would not immediately appeal to heat at depth in the Earth, and would even implicitly reject heat as having any significant role, as Werner did. Werner was extrapolating from observations, but they were local and very limited for such a grand, global interpretation. His globe-encompassing ocean was a conjecture for which there was little direct evidence, and it begged the question of what happened to all that water? We should allow of course that Europeans were pre-conditioned for this kind of idea by the Biblical account of the great flood. That Werner's ideas became so influential is a reflection perhaps of the power of authority and rhetoric, the Bible's and Werner's both, to direct our thinking.

Hutton, in his argument with Werner and the neptunists, did appeal to heat as the primary agent of geological processes. Did Hutton's appeal to heat have any better basis?

I think so. It was not really a novel idea that the Earth's interior is hot. There is supporting evidence available to common experience in some places. Hot magma issues from the earth's interior through volcanoes, and there are hot springs in volcanic areas. Both of these observations were known to European scientists from the volcanos of Italy. It was also becoming known that the temperature in very deep mines is uncomfortably high.

This knowledge is sufficiently widespread that many people have the opposite misconception—that the interior below the crust is molten. It is not uncommon to hear people speak of the earth's molten core, and mean everything below the crust. Geologists through the first half of the twentieth century might have done well to ponder this common belief more deeply, because it is quite a reasonable inference, and it becomes more so in the light of more precise knowledge of temperatures at depth, which we will look at in the next chapter.

First though, there is another implication. Starting in mid-nineteenth century, a different argument about rates and ages raged, this time between geologists and physicists. The physics argument hinged on the Earth's internal heat.

William Thomson, who later became Lord Kelvin, was a brilliant young physicist who did much to establish the science of thermodynamics. Kelvin was disturbed by the claims of Hutton and Lyell that the Earth is in a steady state. The second law of thermodynamics, which Kelvin helped to formulate, requires that heat will tend to dissipate, so any process driven by the Earth's original heat should slow down. To Kelvin, the geologists' claim amounted to a perpetual motion machine, and he vigorously objected.

Observations of the increase of temperature with depth in the crust were sufficiently well-established by the middle of the nineteenth century to be the basis of Kelvin's estimate of the age of the earth. From deep mines it was known that the temperature increases by 15–20 °C per kilometre downwards. This rate of increase with depth is known in the business as the *temperature gradient*.

Kelvin assumed the earth started in a molten state. His argument was not about the time it would take for the Earth to completely solidify. According to the physics Kelvin was appealing to, namely conduction of heat through a solid, that would take hundreds of billions of years, and the temperature gradient would by then be much lower than the presently observed gradient. Rocks are not very good conductors of heat.

Rather, Kelvin asked, how long would it take for the temperature gradient near the surface to decrease to the present value, starting from a molten state? Initially the gradient would be very steep, because molten rock would be right at the surface. If the Earth cooled from the outside by the action of heat conducting from the interior to the surface, the temperature gradient near the surface would decrease with time in a predictable way. Knowing the thermal conductivity of rocks, and estimating the initial temperature as being close to that of molten rock, Kelvin thus calculated the age of the earth to be about 100 million years, with an uncertainty of a factor of about 4.

Now it happened that Kelvin had also estimated the age of the sun, but based on different reasoning. He had assumed that the sun's energy was derived from gravitational energy during its formation, and calculated the time for which it could sustain its present rate of heat loss. The answer he got was also around 100 million years. He was very impressed by the consistency of his ages of the sun and the Earth, which seemed to put the estimates beyond any likelihood either of coincidence or of serious error in the estimates, because the arguments were rather different for each. (As it happens, in the light of our present understanding, I can't think of any reason why it is anything more than a coincidence. There was more going on in both the Earth and the sun than Kelvin was aware of.)

Kelvin published his results, and eventually got the attention of geologists, who were claiming the Earth to be rather older, at least several hundred million years.

Given the uncertainties on both sides, it may seem strange that a heated and prolonged dispute ensued. We must appreciate that Kelvin was arguing not just for a particular number somewhat smaller than the geologists, he was arguing against the extreme position of Hutton, Lyell and others that the Earth was in a steady state of indeterminate age. Kelvin's fundamental point was that his estimate of age yielded a finite number, rather than an infinite or indeterminate number. His perspective was that a steady state Earth contradicted the laws of physics which he had helped to establish.

Evidently Kelvin did not concede that Lyell's Earth was much closer to a reasonable state of slow decline than the catastrophists' Earth that Lyell had been challenging. Even from the modern perspective, remembering that the interpreted geological record at that time was confined to the Phanerozoic, Lyell was essentially correct concerning tectonics: there has been little discernible overall change in modes and rates of tectonics within this period. Unfortunately the argument degenerated into a squabble about whether the age of the Earth was less than or more than one hundred million years. In pursuit of his point, Kelvin through his later life pared his upper limit progressively down, to 24 million years in 1897.

It is salutary to review how much perspectives had changed between 1780 and 1860. Clerics had argued for an age of the order of 10,000 years, and possibly only 6000 years. Early European geologists tried to fit their theories into the biblical time framework, and assumed as well that the more distant past had been far more violent than the more recent past. Hutton and Lyell argued against a short, catastrophist, declining history of the Earth's activity, but made no quantitative estimates. When Darwin finally made an estimate using their methods, he got a number of the order of 100 million years. This was surely sufficiently different from the Biblical estimate to justify the rhetorical steady-state position of Hutton and Lyell. Kelvin in turn reacted against their rhetorical position, justifiably in the light of emerging knowledge of physics, but he argued on the basis of a number that was also of the order of 100 million years.

That little summary is not meant to be too critical of the participants in the debates. Each of the main participants was endeavouring not only to reach their own conclusion, but to defend it from vigorous attack by others. In the heat of those debates, each side tended to be pushed to a greater extreme, the better to distinguish their claim from others. On the other hand Hutton's claim of a steady state was a bold conjecture with little to support it when he first made it. Lyell did not really need to take it as literally as he did, and he did not need to be so rigid in his adherence to it, especially regarding the fossil record that was revealing a major progression of life with ever increasing clarity.

To his credit, Kelvin qualified his estimates, at least early on, with the explicit assumption that there was no unknown physical process at work. In fact there were two. One of these is well-known, the other deserves to be better known.

First, the discovery of radioactivity and nuclear reactions near the end of the nineteenth century revealed energy sources that could sustain the heat of both the Earth and the sun for billions of years. Rutherford was able to proclaim in 1904, in effect, that although Kelvin's number might not have been right, 'the old boy' had never actually been wrong.

The second neglected process is central to the larger story being told here. In 1881 the Reverend Osmond Fisher published a book, entitled *The Physics of the Earth's Crust*, in which he argued that the crust must reside on a plastic (that is, deformable) substratum. We will look at some reasons for this idea in the next chapter. Fisher noted that Kelvin's argument depended on the assumption that heat is *conducted* out of the Earth's interior. Given the known thermal conductivity of rocks, which is quite low, this process could cool the Earth only to a depth of about 100 km in 100 million years. However if the interior were deformable, then convection could transport heat from a much greater depth to replace that lost through the surface. In this way the presently observed temperature gradient in the crust could be maintained for much longer than Kelvin's estimate of 100 million years.

Marcia Bjornerud, in *Timefulness*, notes another factor in the dispute, less complimentary to Kelvin. In his address to the British Association for the Advancement of Science on becoming President in 1872, Kelvin allowed that he never really believed that natural selection could account for evolution, if indeed evolution had occurred, and that 'Overpoweringly strong proofs of intelligent and benevolent design lie around us'. In seeking to discredit the geologists and Darwin he was defending his religious convictions.

Fisher's work was overlooked for a long time in subsequent debates, which is a shame because it could have allowed much more constructive progress, not only on the age of the Earth but on mountain building and geological processes in general. There was not sufficient evidence in Fisher's day to decide such questions, but people might have been more receptive to alternative ideas that ultimately led in the right direction. Of course it is easy to be wise in hindsight.

Even today, no-one has directly measured the age of the Earth. The oldest rocks known are about 4 billion years old, and a few grains of the mineral zircon, incorporated into younger sediment, have ages up to 4.4 billion years. By now we do have a clear record of changes in the way the Earth works, because we can see much further back than Hutton and Lyell could. We even have a 'vestige of a beginning' that allows us to estimate the age of the Earth, but it is a subtle one requiring some assumptions and comparisons with meteorites for its interpretation.

The ages of most meteorites are similar, and have been measured using the decay of two uranium isotopes into lead isotopes, yielding an age of 4.57 billion years. It has also been demonstrated that estimates of the mean lead isotope composition of the Earth are similar to the meteorite lead, which is consistent with the Earth having a similar age. But that is not quite the end of the story, because different assumptions can lead to different estimates of the Earth's mean lead isotope composition, and for all reasonable estimates of this the age obtained is significantly younger than that of the meteorites, say around 4.52 billion years, perhaps 50 million years younger than the meteorites. Further, the event dated in this way need not be the formation of the Earth. More likely it was the separation of the core from the mantle, which may have required Earth to reach a substantial size before gravity did its work. Still, these are details really, the conclusion is that the Earth is around 4.5 billion years old.

In the Judeo-Christian culture the world was presumed to be some thousands of years old. Bishop Ussher proclaimed that the world began in 4004 B.C., based on his scriptural studies. Modern archaeology considers that the first cities were established around that time. It was in cities that writing was developed, at first just to keep accounts of crops and taxes, but with gradually broadening uses like writing out the rules and, eventually, writing down stories like the epic of Gilgamesh. So Bishop Ussher seems to have made a very good estimate of the age of the 'civilised' world, the world of cities.

Because of the authority the written word came to have, when joined with Christianity in the Bible, European geologists had to find their way out of the presumption of a young world before they could accept the implications of their own observations: Charles Lyell said every pebble in the great conglomerate tells its own tale, but someone had to be willing to listen to the tale. The heroic tales of the founding and development of a science of geology were constrained and determined to a considerable extent by the culture they were emerging from.

It is well to realise that people in other cultures did not have the same conceptual constraints, and people in other cultures could be astute observers of their world.

Hindu culture has a conception of the world as flowing in cycles, cycles of creation and destruction, death and rebirth. There are cycles within cycles, and the larger cycles are very large indeed, millions or billions of years in duration according so some interpretations. So Hindu geologists would have been more open to the tales of the pebbles.

If you are inclined to think 'Yes, but Hindus were not smart enough to invent geology' you should be aware that Europeans had to learn from elsewhere how to do arithmetic, not so long ago. Venetian traders learnt of the decimal notation of numbers from the Arabic world, and the Arabs may well have learnt it from Hindus. European science could not have developed without it, nor European and world commerce. The story is well told by Jane Gleeson-White in *Double Entry* (2011). Before Arabic/Hindu numerals, Europeans had to write out their arithmetic problem in clumsy Roman numerals, then do their calculation on an abacus, then write the result down in more clumsy Roman numerals. With the new numeral system, the entire problem could be worked out on the page, visible and preserved. It was a miraculous development. Algebra had already been developed. Double-entry bookkeeping followed, and that transformed commerce. European culture had its time of blossoming, but so had others before it.

Australian First Peoples' many cultures have stories of the rainbow serpent travelling the land and forming many of its features, 'when the world was soft' as one story teller put it. One story has it that when the rainbow serpent travelled the route that became Australia's largest river, Dhungalla or the Murray, it cut south within Victoria and followed the present Glenelg River to the Southern Ocean on the border with South Australia. Only later did the Rainbow serpent take a new route further west into South Australia before turning south. It happens that geologists agree that the Murray used to connect with the present Glenelg River, but a modest uplift of the land re-routed it further west.

You can make of that story what you will. Some say it shows people were in Australia even long before the 65,000 years presently documented, but that would require them to have been here for millions of years. My supposition is that someone recognised and interpreted the geomorphology, seeing an older river valley still imprinted in the landscape. There is another place near Echuca where a more recent uplift of ten metres or so deflected the river, both north and south—the land is very flat in that region. I was born next to the northern anabranch so formed. I am told you can clearly see the old river channel continuing over the uplift. We know First Australians were and are astute observers of their environment, knowing virtually every species, and its place in the order of things, within their Country.

Chapter 5
Yielding Rocks

Shashi Tharoor does not think the British presence in India was of net benefit to Indians. He explains at some length and rather persuasively, in his book *Inglorious Empire: What the British Did to India*, that the British effort was focussed on extracting wealth, and any benefit to the locals was incidental and marginal. Railways, a frequently-cited example, were focussed literally on extracting wealth for export, rather than on averting famine or helping Indians to get around their own country. The British depredations ruined the jewell of the mediaeval world.

There were also a few incidental benefits of the British occupation to general scholarship and knowledge. A well-known one was the realisation that Sanskrit and Latin are related languages. A lesser-known bit of technical work was the detection of a deficit of gravitational attraction from the Himalaya mountains.

Even though gravity is a very weak force, there is enough mass in large mountain ranges to exert a measurable attraction. This was enough to cause a discrepancy between two different surveying methods near the Himalayan range. Reginald Daly gave a good account in his *Strength and Structure of the Earth* in 1940. One surveying method was based on triangulation and the other on astronomical sighting. The astronomical sightings were done relative to the local vertical, as determined by a plumb line, which is a lead weight hanging on a thread. The attraction of the mountains pulled the lead weight a little bit towards them, so the plumb line was not showing the true vertical. The surveys were done by Sir George Everest, and later the highest mountain was given his name, even though it had perfectly good names already: Sagarmatha (Nepali) and Chomolungma (Tibetan).

In 1855 J. H. Pratt did some calculations to demonstrate the effect. However his calculations revealed that the attraction was only about a third of what would be expected from the visible mountain range. This was not the first time such a deficit had been found. A deficit was first recorded for the Andes mountains by Pierre Bouguer, on an expedition between 1735 and 1745, and another was found in 1849 by Petit near the Pyrenees. It was the analogous observations from Everest's surveying in India that led to further hypotheses. Explanations for the deficits were offered by Pratt (1859) and by G. B. Airy (1855).

G. F. Davies, *Stories from the Deep Earth*,
https://doi.org/10.1007/978-3-030-91359-5_5

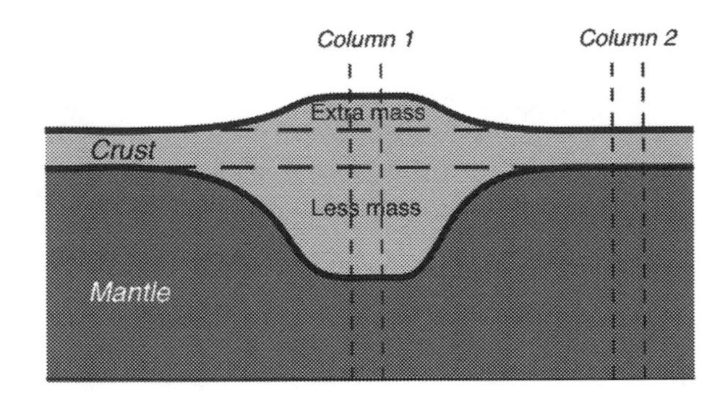

Fig. 5.1 Airy's conception of mountains or a plateau with a *root* comprising thickened crust. The crust is assumed to be less dense than the substratum (the mantle), so it exerts an upward buoyancy force that tends to balance the downward force of the weight of the mountains

Every geology student learns about about Pratt's and Airy's variations on how to account for a density deficit under mountain ranges. What I had never appreciated until I read Airy's short paper, extensively quoted by Daly, was how penetrating and far-reaching was Airy's thinking. He is famous for hypothesising what was later called the condition of isostatic balance or isostasy, but his thinking went to the core of the subject of tectonic mechanism.

Both Pratt and Airy proposed that there must be low-density rock below the mountains, to partially balance the extra weight of the mountains. The lower density would add up to less mass in total, a mass deficit. Pratt suggested that such differences in density might have arisen when the Earth was young and liquid, for example by differential thermal expansion or contraction, but he was not very specific about how this came about.

Rather than assuming that density distributions had been fixed at the beginning, Airy thought the Earth must have been subject to continuing 'disturbing causes' through its history which could change not only the topography but the densities under the surface. He noted that the shape of the solid part of the Earth (excluding mountains) closely approximates the shape of the liquid ocean surface and this had been taken by physicists to indicate 'either that the interior of the Earth is now fluid or that it was fluid when the mountains took their present forms'.

Airy therefore assumed an outer, non-deforming 'crust' and a denser, fluid interior. He argued first that a broad plateau could not be supported alone by the strength of the crust, demonstrating that the leverage required at its edges required a tensile strength that was very implausible, given that the crust is known to be riven with fractures, even if the crust were 100 miles (160 km) thick. He then asked how else such plateaus might be supported, and answered himself.

> I conceive there can be no other support than that arising from the downward projection of a portion of the earth's light crust into the dense [substratum]; ... the depth of its projection downwards being such that the increased power of floatation thus gained is roughly equal to the increase of weight above from the prominence of the [plateau].

He compared the crust to a raft of logs floating on water, wherein a log whose top is higher than the others will be correctly inferred to be larger and thus to project deeper into the water than the others. His idea is sketched in Fig. 5.1.

The 'downward projection' of crust has become known as the *root* of a mountain or plateau. The 'power of flotation', or buoyancy, of the root will balance the extra weight of the plateau if the mass of rock in Column 1 in the diagram is the same as the mass of rock in Column 2, away from the plateau. It is just an application of Archimedes' principle. This condition of balance is known in geology as an *isostatic balance*, or briefly as *isostasy*.

Airy then showed how the root will reduce the net gravitational attraction, countering the extra attraction of the plateau. At a distance large compared with the depth of the projection the net gravitational perturbation will approach zero. He noted that one would not expect that there would everywhere be a perfect isostatic balance, but that the strength of the crust would allow some mountains to project higher or some roots to project deeper than in the isostatic condition. This would be especially true of mountains of small horizontal extent, since the leverage required to hold them up is smaller.

Airy gave a second reason for thinking the substratum would behave as a fluid. Having noted how the shape of the Earth implies a fluid interior, he goes on.

> This fluidity may be very imperfect; it may be mere viscidity; *it may even be little more than that degree of yielding which (as is well known to miners) shows itself by changes in the floors of subterraneous chambers at a great depth* when their width exceeds 20 or 30 feet [7 or 10 meters]; and this degree of yielding may be sufficient for my present explanation. [Emphasis added.]

Here, very clearly, is a concept of a solid, rocky, but deformable interior based on the direct experience of miners, for whom it is a matter of life and death to take it properly into account. Perhaps because mining was considered a trade rather than a science, this telling observation did not gain as much currency as it deserved.

Airy's presumption that 'disturbance' was not confined to the formation of the Earth but continued through Earth history gained important support from ongoing geological observations (and may have been motivated by them). In 1859 J. Hall presented evidence from New York State of slow, continuous adjustment of the Earth's surface to changing loads, by demonstrating that sediments now buried deep in thick sedimentary sequences had been deposited in shallow water. Lyell had made a similar observation in the Paris Basin and there have been many others since. The implication is that the sediments were not deposited into a pre-existing deep depression, but that the depression deepened as the sediments accumulated. This requires that a substratum at depth yields enough to be displace out of the way of the sinking crust.

By 1889 there was accumulating evidence that broad crustal structures like plateaus are close to isostatic equilibrium, whereas the crust has enough strength to hold narrower structures like single mountains out of equilibrium. C. E. Dutton in the United States formalised the idea and proposed the name *isostasy* (Greek: isos, equal; statikos, stable). Dutton actually preferred the term isobary, or equal pressure, but this was already in use in another context.

F. R. Helmert in Germany conceived in 1909 that the depth of the compensating root could be constrained by the form and magnitude of the gravity anomaly at

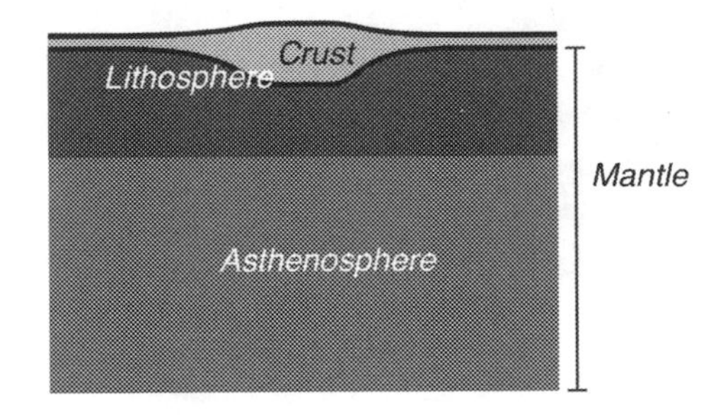

Fig. 5.2 The crust, less dense and compositionally different; the mantle, denser than the crust; the lithosphere, stronger because it is colder; and the asthenosphere, in the solid state but capable of being slowly deformed like a fluid

the edge of a broad structure, and he and others used observations near continental margins to deduce that density anomalies extend to depths of around 100 kms. This implied that the strong, non-fluid 'crust' must extend to such depths, in order for the density anomalies to persist. This gave rise to the idea of a strong, non-fluid *lithosphere*, being the cold outer part of the Earth that is stronger because (presumably) it is colder.

At around the same time the Croatian seismologist Andrija Mohorovičić discovered a seismic discontinuity, typically at a depth of 35–40 km in continental areas, due to the velocity of seismic waves being greater below the discontinuity. This boundary is now known as the Mohorovičić discontinuity, or Moho, and it defines the crust, the upper layer of lower density and thus of different composition from the underlying material. Thus the crust is part of the lithosphere, which is thicker.

In 1914 J. Barrell in the United States proposed the term *asthenosphere* (weak layer) for the deformable region below the lithosphere. Barrell was willing to assume that the thickness of the asthenosphere is as great as 600 km, in order to reduce the amount of deformation required to accommodate surface uplifts. This allowed him to argue that a deformable asthenosphere was not incompatible with it being in the solid state, as shown by its ability to propagate seismic shear waves.

Thus by 1914 there was a clear picture, well based on observations, of a lithosphere about 100 km thick and strong enough, on geological time scales, to support topography up to a width of about 100 km. Topography on broader scales was known to be approximately in isostatic balance. That included the topography of continents relative to the sea floor. It was inferred that the balance occurs because the asthenosphere, below the lithosphere, behaves like a fluid on geological time scales, in spite of being in the solid state. Figure 5.2 illustrates the distinctions among crust, mantle, lithosphere and asthenosphere. Crust and mantle have different *compositions*. Lithosphere and asthenosphere have different *strengths*. The crust is actually part of the lithosphere. The non-crust part of the lithosphere and the asthenosphere are both part of the mantle, the denser rocks underlying the crust.

A different kind of observation was developed through this period that strongly supported this picture. However there were two other observations that complicated it.

The supporting observation was that the land around the Baltic Sea was slowly rising, leaving a series of elevated old shorelines on surrounding hillsides, the higher

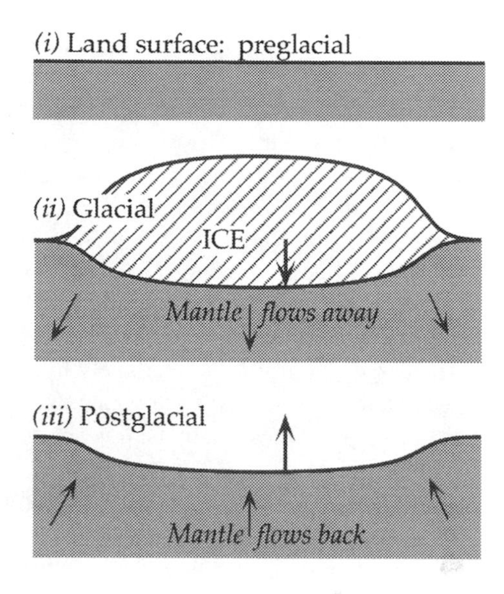

Fig. 5.3 Postglacial 'rebound'. During an ice age the weight of a thick ice sheet depresses the surface, which in turn forces deeper mantle material down and away. The process is reversed after the ice sheet melts: the depression left by the ice slowly rises back to its previous level. (Because the Fennoscandian ice sheet was about 1000 km wide, the strength of the lithosphere provides little resistance, and it is not shown in the sketches). From Davies [4], reproduced with permission

ones evidently being older. Even in 1865 T. F. Jamieson argued that this could be explained as a delayed response to the removal of a thick ice sheet after the recent ice age. (The ice sheet is called the Fennoscandian ice sheet, named for Finland and the Scandinavian countries around the Baltic Sea.) Jamieson supposed the weight of the ice sheet would have forced the Earth's surface downwards, and this in turn would force deeper mantle material down and to the sides. The process would be reversed after the icecap melted, creating a delayed 'rebound' of the surface. The idea is illustrated in Fig. 5.3.

If the mantle were simply solid and elastic, there would be no delay in the process. The depression would form as the ice accumulated and it would rebound as the ice melted. On the day the last ice melted the depression would be gone. The fact that the depression had still not gone away 10,000 years after the ice melted implied that the mantle is deforming like a fluid.

A *viscous* fluid, like honey, is one that resists changes in its shape but that ultimately does yield to whatever force is acting on it. Technically, the *rate* of flow is proportional to the applied force. The amount of resistance is characterised by the fluid's *viscosity*: cold honey, which flows slowly, is more viscous than warm honey, which flows faster.

By the 1930s well-founded estimates of the viscosity of the asthenosphere had been derived from this logic by several workers. By then there were accurate dates of the old shorelines on the hillsides, going back about 10,000 years, the oldest being the highest. In that time the land surface had risen about 250 m. The most relevant calculation for our purposes was by N. A. Haskell. He assumed the asthenosphere extended deep into the Earth and obtained a viscosity of 3×10^{20} Pa s (the unit of viscosity is Pascal-seconds). Yes that is a very large number, 3 followed by twenty zeros. Even under the weight of several kilometres of ice the mantle takes 10,000 years or more to flow out of the way.

But 10,000 years is not long geologically speaking. The observations of postglacial rebound, as it is known, thus supported the inferences from subsiding sedimentary basins, and from deep mines, that the mantle behaves like a fluid in geological time. The fluidity, or *viscidity* as Airy called it, had even yielded a quantitative measure of mantle viscosity.

Now to the complications in this picture. We have already encountered the first of these, which is that some earthquakes occur down to depths of nearly 700 km. The second complication was that even on the largest scale the Earth is not *quite* in hydrostatic equilibrium. A completely hydrostatic Earth should have an equatorial bulge due to rotation, but it was found that the equator bulges by about 20 m more than this, and that the equator itself is not uniform, bulging more in some longitudes than in others. Significant stress is required to support these bulges.

The shape of the Earth had been determined by French surveys in the eighteenth century, but it was Harold Jeffreys, in his book *The Earth,* first published in 1924, who demonstrated that the bulges are slightly larger than can be accounted for by the Earth's rotation. It was known by then that the Earth's rotation was faster in the past, and Jeffreys supposed that the bulge was a 'fossil', reflecting that faster rotation in the past. This implies that the mantle is very slow to adjust and hence very strong or resistant to flow. It implies a viscosity of at least 10^{26} Pa s. The occurrence of deep earthquakes was taken by Jeffreys as corroboration of this interpretation. Evidently Benioff followed Jeffreys' conclusions.

However there are alternative interpretations of both observations. We have already seen the alternative for deep earthquakes: that they occur within the cold, stronger slab of lithosphere sinking under a subduction zone, as depicted in Fig. 3.1. It is still not quite clear how those deep earthquakes are generated, but it is clear from the new picture that the deep earthquakes are occurring in a zone that is not typical of the mantle at those depths, of a few hundred kilometres. In this interpretation the deep earthquakes are not inconsistent with the other evidence for a flowing asthenosphere.

An alternative explanation of the excess bulges is that they are supported by stresses in a fluid, convecting mantle. There had been occasional suggestions of convection in the mantle, including by Fisher and Holmes already mentioned. However few seemed willing to seriously advocate it until Keith Runcorn in 1962, motivated by geomagnetic observations supporting continental drift that we will come to later. In 1969 Peter Goldreich and Alar Toomre showed that internal fluid motions were capable of explaining the excess bulge. They made two other key points as well. First, a viscous Earth would slowly shift on its axis so as to bring any bulges to the equator. Second, the bulge around the equator is not uniform, it actually varies around the equator by about as much as the variation from pole to equator. Since variations around the equator cannot be explained by faster rotation in the past, the bulges must be due to another cause. With these arguments Goldreich and Toomre showed that the viscosity of the mantle cannot be more than about 3×10^{20} Pa s, very consistent with the viscosity inferred from postglacial rebound.

What about the other lines of evidence for a deformable mantle: isostatic balance and slow subsidence of sedimentary basins? Jeffreys argued that the approximate isostatic balance of mountain ranges was due to the fracturing of the crust by tectonic forces, and subsequently by secondary gravitational (buoyancy) forces induced by the (supposed) resulting topography. He drew attention to the distinction between the strength of un-fractured rock and the much lower strength of fractured rock. He supposed that it was the tectonic forces that first fractured the rock, and that the strength implied by remaining isostatic imbalance is a measure of the strength of fractured rock.

Reginald Daly at Harvard University in the US disputed Jeffreys, at Cambridge UK, on several grounds. He pointed out that Jeffreys' hypothesis could not account for slow isostatic adjustment away from mountain belts in response to erosion and sedimentation, nor for observed continuing adjustment to deglaciation. Daly also noted experiments by his colleague P. W. Bridgman that had shown that fractures healed quickly at high pressures. As well, we can note the internal contradiction in Jeffreys' argument that the remaining isostatic imbalance should still have reflected the strength of un-fractured rock: any un-fractured parts could still be out of equilibrium, and it would have been necessary to overcome the un-fractured strength in order to bring them closer to balance.

Daly's ideas, presented in his book *Strength and Structure of the Earth*, deserve more recognition. His thinking was wide-ranging and adventurous, and he came remarkably close to some modern concepts, as we will get to shortly, based on his understanding of the lithosphere and asthenosphere. His willingness to think about a deformable mantle led him to resolutions of puzzles that were more integrated and more enduring. On the other hand Jeffreys' proposals were more piecemeal, still somewhat speculative and not all contradictions were resolved.

Not all of Daly's ideas were well-based. For example he regarded the asthenosphere as being in a vitreous (glassy) rather than a crystalline state, despite his evident awareness of Airy's point that crystalline rocks were known to deform in deep mines, and despite his colleague David Griggs' experiments on rock deformation. He proposed to explain the large-scale bulges of the Earth by supposing that below the asthenosphere is a 'mesosphere' of greater strength, though this neglects to explain how stresses maintained in the a strong mesosphere would be transmitted through the asthenosphere to the surface.

Nevertheless, acknowledging that he was indulging in conjecture, Daly offered several suggestions to explain the occurrence of deep earthquakes. He proposed that the asthenosphere is heterogeneous, being strong enough to bear brittle fracture in some places. He proposed ways that this might come about, the most interesting being that blocks of lithosphere might founder and sink through the asthenosphere. Furthermore, noting that suddenly imposed stresses might induce fracture even in the deformable asthenosphere, he suggested that pressure-induced phase transformations, of the kind recently observed by his colleague Bridgman, in such sinking blocks might be a suitable trigger. This is an idea still very seriously entertained.

Daly proposed that the foundered or 'stoped' lithospheric blocks could plausibly originate during compressional mountain building:

> mountain making of the Alpine type seems necessarily accompanied by the diving of enormous masses of simatic [mantle], lithospheric rock into the asthenosphere. Thus the belt under the growing mountain chain is chilled by huge, downwardly-directed prongs of the lithosphere, as well as by down-stoped blocks.

He noted that this would explain the occurrence of deep earthquakes 'under broad belts of recent, energetic orogeny'. This picture of the lithosphere, including 'prongs' projecting down under zones of compression, is remarkably close to the modern picture of a subduction zone in Fig. 3.1.

The whole pre-war geology department at Harvard University is worthy of comment. Percy W. Bridgman was pioneering high-pressure experiments examining compressibility of rocks and minerals, and the pressure-induced changes of crystal structure they often undergo. David Griggs was measuring the strength of rocks, including their brittle fracture and the effect of pressure on brittleness. Griggs also proposed mantle convection as a mechanism of continental drift and mountain building. Reginald Daly was synthesising broad aspects of physical geology. They were joined by a young Francis Birch, who set about measuring elastic properties of rocks and minerals under high pressure, which allowed the interpretation of seismic velocities in the Earth's interior. Birch resumed that work after the war, and he was the last of that group still active there when I arrived there as a green postdoctoral fellow in 1973. (David Griggs had moved to UCLA postwar.)

The US had not attained the academic prominence, and dominance, that it developed after the war, but the work at Harvard was visionary, and it built fundamental understandings of the effects of temperature and pressure on the material properties of minerals. The nearest equivalent in the UK seems to have been Arthur Holmes at Newcastle-upon-Tyne. Holmes had pioneered the use of radioactivity to date rocks, which became a quite fundamental tool in geology. That story is well told in Cherry Lewis' *The Dating Game* (2000). He was also a wide-ranging integrator of physical geology, like Daly. On the other hand Jeffreys at Cambridge was a very good mathematical physicist who did fundamental work on the subtleties of the Earth's rotation, which are complicated, but he had less understanding of either physical geology or material properties. The social reality was that Jeffreys had the plum job in the UK and Holmes had to be content in a provincial university. Harvard also was still somewhat on the outer, pre-war, relative to European science.

To summarise the evidence for a creeping mantle, gravity measurements established that mountains are close to an isostatic balance, and observations of associated sedimentary sequences showed that there are slow and continuous adjustments of the Earth's surface to changing loads. Observations of post-glacial rebound of the Earth's surface supported this inference and yielded quantitative estimates of the viscosity of the mantle. Observation of non-hydrostatic bulges were at first taken as evidence for a rigid interior, but were later re-interpreted as indicating a fluid interior with a viscosity comparable to that inferred from post-glacial rebound. Deep earthquakes remained a puzzle, but Daly conjectured that the asthenosphere in which they occur is abnormal, and that the abnormalities might be associated with the active mountain belts that overlie them.

So far we have looked at evidence that the mantle is deformable using observations of the Earth, but there is also information from laboratory experiments. Cold rock does not deform significantly, even on geological timescales. It is only at high temperature that some slow deformation may occur. So first let us look again at the evidence that the mantle is indeed hot.

Many measurements have been made of temperature in mines and boreholes, and it has been long established that the temperature in the crust increases with depth at a rate usually in the range 15–25 °C/km (degrees centigrade per kilometre). This directly implies quite high temperatures, even in the crust. Continental crust is typically 35–40 km thick, so by extrapolating the observed gradient we can estimate temperatures at its base to be in the range 500–1000 °C.

Granites melt at temperatures of about 750 °C. The rocks of the mantle, below the crust, begin to melt at about 1200 °C. This makes it plausible that the deep crust and the top of the mantle sometimes melt to yield volcanic magmas. It would also be plausible, in the absence of contrary evidence, that the Earth's interior is entirely molten below a depth of about 80 km. However we know the mantle is not entirely molten because it transmits both kinds of seismic waves, compressional and shear. If it were molten it would not be able to transmit shear waves.

Estimating the temperature at depth is not quite so simple as just presented. This is because of a small amount of radioactivity in the upper parts of the continental crust. The main radioactive elements are uranium, thorium and potassium. In most places the amount of radioactivity is not enough to cause us much harm. If that were not so the living world would not have been able to develop.

I will briefly explain what difference this makes, but the details are not too important here, it just yields an interesting number. So, the radioactivity generates some heat, and a significant fraction of the heat emerging from the continental crust is due to this radioactivity. This implies that less of the heat is coming from the mantle, which implies that the temperature gradient in the lower crust is smaller than we see near the surface. From this we can estimate temperatures at the base of the continental crust are likely to be in the range 400–700 °C, and the depth at which the mantle would start to melt is about 100 km. That is about the same as the estimates of the thickness of the lithosphere mentioned earlier.

What if the temperature gradient in the crust, let's say 10 °C/km, continued unchanged down through the mantle? The mantle is nearly 3000 km thick, so at the bottom it would be around 30,000 °C. It would not just be melted, it would be vaporised. But we still know the bottom of the mantle is solid, so something else must be happening. The something else is convection in the mantle, which slowly moves material up and down, carrying heat with it. Heat travels through the mantle by being carried, or *advected*, not by being conducted. This means the temperature gradient is much lower than it is in the lithosphere. There is still a small temperature gradient because of the effect of the increasing pressure with depth; this is called the *adiabatic gradient*. Thus the temperature at the bottom of the mantle is around 2,500 °C, not 30,000 °C.

Figure 5.4 shows a schematic temperature profile down through the mantle, along with a schematic of the melting temperature. The melting temperature also increases

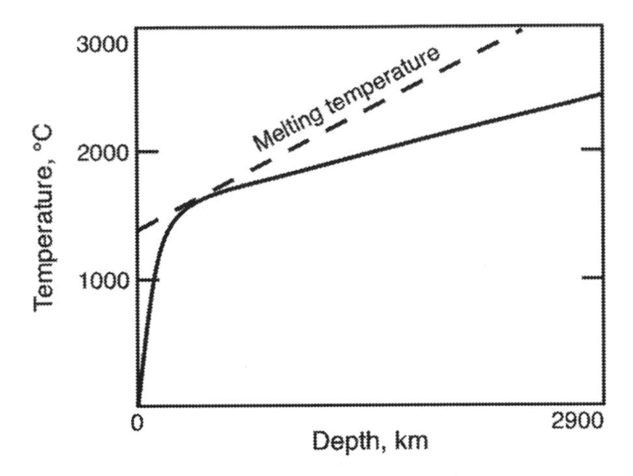

Fig. 5.4 Schematic temperature profile through the mantle. The temperature gradient is steep near the surface where heat is *conducted* upwards. It is shallower at greater depth because heat is carried by *convection*. The small gradient, called the *adiabatic gradient*, is caused by the increasing pressure. A schematic melting temperature is included. The mantle temperature is close to the melting temperature around 100–200 km depth

at higher pressures, but faster than the mantle temperature. The result, you can see, is that the mantle temperature is close to the melting temperature around 100–200 km depth but further from it above and below.

It is this close approach of the mantle temperature to the melting curve that accounts for volcanos. If some portion of the mantle is carried upwards, say under a mid-ocean rise, it will cross the melting curve, generating magma that may be erupted at the surface. For a different reason there are also volcanos at subduction zones: *water* reduces the melting temperature of the mantle rock. The subducting lithospheric slab carries some chemically bound water down with it. High pressure causes that water to be released from the slab at around 100 km depth, and the result is a different kind of volcano at subduction zones. A third kind of volcano is due to mantle plumes, which we will come to later.

The fact that the mantle approaches the melting curve is probably not coincidence. The mantle was probably hotter in the past. This would have produced more volcanism, which would have helped to reduce the temperature of the mantle. There is another reason as well, but the net effect is that the mantle was cooled until it was only melting a little bit, and then the cooling slowed. At some time in the far future it may cool enough that there will be no melting and no volcanos.

Most of us know, or have heard, that if a blacksmith heats an iron horseshoe until it glows red hot then he can shape the shoe to fit the horse's hoof. As I noted earlier, most of us are not so familiar with the idea that rocks can be malleable.

If a rock is cold, it is brittle, and if it is heated sufficiently it melts and becomes magma. There is however a range of intermediate behaviour, at temperatures near but below the melting temperature, in which rocks and other solids can deform without breaking, but this behaviour is only rarely perceptible in non-metals. This is because the deformation is slow, and we require an elapse of time before we can see it. An example commonly cited is that of glass, which at normal temperatures deforms under the action of its own weight at a rate that makes its deformation observable over decades or centuries. Thus it is reported that drink bottles recovered from the desert have sagged like an object in a surrealist painting, and that windows in old cathedrals of Europe are noticeably thicker at the bottom than at the top because they have sagged slightly.

For solids to deform without breaking also requires that they are not stressed too greatly. It turns out that the action of pressure suppresses brittle fracture in favour of ductile deformation. In effect, the two sides of a potential fracture are locked together under the action of pressure, and then the solid's only available response is its tendency to deform slowly throughout its volume. Of course the pressures are very high in the Earth's interior, and evidently sufficient to suppress brittle behaviour completely at depths greater than about 100 km.

The result is that with the high temperatures and pressures of the Earth's interior, and enough time, rocks can deform and flow, and thus be considered as fluids, even though they are solid in practical experience. The rate of deformation is extremely slow: about twenty orders of magnitude slower than liquids of common experience, for similar stress levels. Seismic waves have periods in the range of seconds to thousands of seconds, and the deformation is so small in this period that shear waves can be transmitted with little dissipation of energy. Thus to seismic waves the material is effectively a solid. On the other hand, there can be significant deformation in a few thousand years.

It also turns out that the viscosity of mantle rocks is strongly dependent on temperature. As the melting temperature of the rock is approached, the viscosity drops by nearly a factor of ten for every hundred degree rise in temperature. This gives us a simple and robust argument that the mantle is likely to be convecting. It was reportedly noted by Harold Urey but only quantified by D. C. Tozer in 1965. In 1980 two groups, one including me, realised it also allowed us to calculate how the mantle has slowly cooled through the aeons. That is a subject for later.

For the moment the lesson is that detailed understanding of rock deformation is now consistent with the observations of the Earth that imply the mantle must be deformable on geological timescales. At the short timescales of seismic waves, seconds to some minutes, the deformation is so slow as to be negligible, so the mantle manifests as a solid. However at timescales of thousands or millions of years the mantle deforms like a fluid.

Chapter 6
Vagrant Continents

The idea that parts of the Earth's surface have moved slowly over distances comparable to the size of the globe belongs mostly to the twentieth century. Earlier observations that provided evidence for a deformable interior, covered in the previous chapter, concerned motions that are mainly vertical.

The similarity of the South American and African coastlines was remarked on by various writers, even by Francis Bacon in 1620. However Bacon did not suggest movement, according to Anthony Hallam in *A Revolution in the Earth Sciences* (1973). It was perhaps Antonio Snider-Pellegrini in 1858 who first suggested that those continents had been torn apart in an early episode of a catastrophist account of the world. A more comprehensive and gradualist proposal was made by the American F. B. Taylor in 1910. Taylor was stimulated by Eduard Suess' descriptions of the great mountain belts in his five-volume *The Face of the Earth* of 1904–9. Taylor's scheme involved broad motions mainly towards the equator, but with obvious deviations. He did not provide any clear mechanism for these movements.

It was Alfred Wegener who first started to assemble detailed geological and geophysical arguments that the continents had moved. In 1912 he first spoke publicly and wrote of his idea that whole continents had undergone large, slow displacements. These ideas were published in book form in 1915 as *Die Entstehung der Kontinente und Ozeane* (*The Origin of Continents and Oceans*), which went through several editions both before and after his death.

A couple of factors may have facilitated Wegener's boldness of thinking. One was that he was not a geologist by training, but a meteorologist, so he had no particular commitment to prevailing ideas in the geological community. The other was that in Germany at the time the geophysics community embraced meteorology and climatology as well as the 'solid' Earth, which perhaps made their thinking more open to mobilist ideas. In Britain, by contrast, geophysics had emerged from astronomy, as some astronomers turned their attention to the orbital motions and rotations of the planets.

The seed of Wegener's theory came from the similarity in map view of the shapes of the *continental margins* on either side of the Atlantic Ocean. Continental margins

G. F. Davies, *Stories from the Deep Earth*,
https://doi.org/10.1007/978-3-030-91359-5_6

are the geological edges of continents, where their surface drops more steeply to the great ocean depths of several kilometres. The oceans lap onto the continents a little bit, so the coastline reflects vagaries of minor topography. The similarity of coastlines had been remarked upon previously, but the similarity of continental margins is greater.

Wegener's ideas became more definite when he learned of similarities in fossils occurring on opposite sides of the Atlantic, and later of geological similarities. He developed his ideas into the proposal that all of the continents had been grouped into one supercontinent, and that this had fragmented and the pieces had drifted apart starting in the Mesozoic era, about 250 million years ago.

In later editions he added more data, and also used evidence from various kinds of deposits that could be used to infer climate, and thus to distinguish equatorial and polar regions, arguing that the distributions of paleoclimate made more sense if the continents had moved (Fig. 6.1).

The similarities in fossils across the Atlantic were also noted by others, and had led palaeontologists to postulate the past existence of land connections between the widely separated continents. Initially these connections were assumed to be continents that had later subsided under the ocean. Geologists had for decades held that continents episodically emerged from and subsided into the ocean, on the basis of the widespread occurrence of marine fossils in continental deposits. Little was known of the nature of the rocks of the sea floor at this time, which limited the possibility of direct geological tests of the idea.

Wegener argued that such large vertical movements of continents were not viable, because gravity measurements had shown that the Earth's crust is close to isostatic equilibrium, as we saw in the previous chapter. The continents stand higher than the seafloor because continental crust must be less dense, and the continents in effect 'float' in a denser substratum. If large continental blocks subsided by several kilometres, there would be large negative gravity anomalies created, and such large anomalies were excluded by the observations. One response to Wegener's isostasy argument was to assume that the land connections were smaller than continental scale—narrow land 'bridges'. This proposal was *ad hoc*, and has never had any evidence to support it.

Another argument Wegener made against the idea of rising and sinking continents was that if it occurred erratically in space and time, as the geological record suggested, one would expect the elevations of the Earth's surface to be spread more evenly than they are between the highest and the lowest, with a preponderance of areas at intermediate elevations. Instead, the observation is that there are two preponderant levels of the Earth's surface, one at the level of the deep sea floor and one at the level of the continental surfaces, just above sea level.

This *bimodal* distribution of elevation had been recognised for a long time as a first-order feature of the Earth requiring explanation. Wegener's point is a very sound one, that the observed topography looks like a very improbable consequence of the idea of rising and sinking continents. However this argument had the rhetorical weakness that he did not have an explanation for the bimodal topography either. In the absence of explicit mechanisms for either vertical or horizontal displacements of

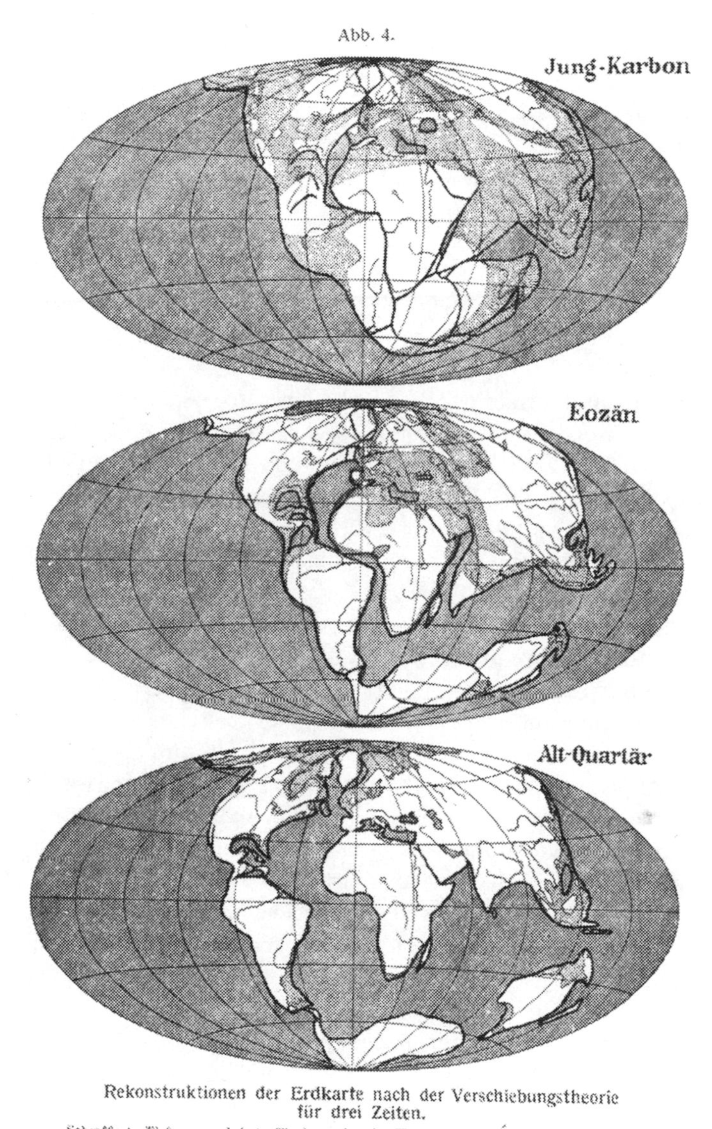

Fig. 6.1 Wegener's 1912 sketch map of the drifting continents, starting from a single continent, Pangea, in the late Palaeozoic, before about 250 million years ago. Reproduction courtesy of Wikipedia [5]

continents, it was not possible to quantify the argument at all, and so the opponents of his theory were freer to make suppositions to suit their point of view, and to note, for instance, that the particular Gaussian distribution that he assumed to make his point was no more probable, a priori, than the observed distribution.

Wegener was tentative about what force or forces might cause continents to drift, writing 'The Newton of drift theory has not yet appeared …' and conceding that it might be a long time before this was clarified. In what can be seen in retrospect as a key tactical error, he suggested a differential rotational force and tidal forces as possible causes. Recall that Darwin had also made a tactical 'error' when he made a rough estimate of the age of an erosional episode, thereby providing a target for Kelvin to snipe at.

By the time of his third edition (1922), which became better known in the English-speaking world, Wegener's theory began to generate strong opposition. According to Bill Menard in *The Ocean of Truth*, there were two reasons in particular that might have contributed to this. One was that Wegener (perhaps like most of us) had begun with the naive idea that his theory was so obvious that it would quickly be accepted. When this did not occur, and he observed palaeontologists failing to understand his argument against land bridges, he became more of an advocate.

The other reason was that he believed that geodetic measurements had revealed Greenland shifting relative to Europe, and that this was a dramatic confirmation of his theory. (The drift rate implied by the data was meters per year, but he had also correlated Pleistocene glacial moraines across the Atlantic that would have required this rate.) The data later turned out to be in error, but by this time he may have become convinced, and adopted a more evangelical approach. When some (but by no means all) of his arguments were found wanting (such as the correlation of moraines), his credibility dropped, and the annoyance of his detractors rose.

Prominent among the opponents of continental drift was Harold Jeffreys. In his 1926 book *The Earth* Jeffreys showed that Wegener's proposed driving forces were many orders of magnitude smaller than would be required to overcome the resistance from the oceanic crust through which the continents were presumed to move. Even in his 1926 edition Jeffreys' language reveals a reaction to Wegener's fervour. He parodied Wegener by accusing him of arguing that a small force acting for a very long time could overcome a much larger force acting for the same time, and characterised this idea as 'a very dangerous one, liable to lead to serious error'. Serious error perhaps, but *dangerous*? Scientists not uncommonly resort to such dramatic language in the heat of argument. The main thing endangered is probably someone's reputation, should an opponent turn out to be right.

Jeffreys' dismissal of Wegener's proposed mechanisms extended to the whole idea of drifting continents. Jeffreys' language reveals that Wegener's proposed forces merely provided a convenient weakness through which to attack the larger theory. It is not clear that Jeffreys made a serious attempt to appreciate Wegener's many geological arguments. He even attacked Wegener's geophysical argument against land bridges, a subject in which he should have been expert, but in which his arguments were inconsistent with the well-known observational basis of isostasy, as we saw in the previous chapter. Thus we see again the process of alternating over-reactions generating a heated scientific debate, just as in the nineteenth century arguments over the age of the Earth, and in many subsequent topics in many areas. Jeffreys remained opposed to continental drift, and later plate tectonics, his entire 98-year-long life.

A primary source of opposition to Wegener's theory was the well-established view amongst geologists and geophysicists that continents are fixed in a strong outer shell of the Earth, the lithosphere. Compressive mountain building forces were supposed to derive from cooling and contraction of the Earth, which generated compressive stresses and occasional failure in this shell. This theory in turn was attracting opposition, because it was not clear it could provide for sufficient crustal shortening to account for the major compressional mountain ranges.

Superficially the model of a cooling, contracting Earth seems attractive, and very compatible with Kelvin's concept, upon which his cooling age of the Earth was based. However if the Earth cools from the outside then the thermal contraction accompanying cooling obviously is confined to the cooled layer, while there would be no change to the hot interior. This would put the outer layer in tension, not in compression. The frequently used analogy of the wrinkled skin of a baked apple does not fit, because the apple shrinks from the inside (due to dehydration) while the skin remains the same. Thus the apple skin is compressed and wrinkled to fit around the smaller interior. In contrast, the Earth's skin would shrink but the interior would not, so the smaller skin would be stretched tighter, or would fault in tension. The great compressive mountain ranges do not fit this picture.

Anyway, Wegener's theory attracted substantial opposition. Wegener's primary activities and reputation during his lifetime were as a meteorologist and polar explorer, and he made several expeditions to Greenland. Unfortunately he perished on the Greenland ice cap in 1930, and thereafter advocacy of his continental drift theory languished. For the next several decades it was not very respectable to advocate anything related to continental drift. It is worth noting that Beno Gutenberg, Director of the Seismological Laboratory at Caltech, in an article written initially in the 1930s but not published until 1951, reviewed an extensive literature of speculative tectonic theories, with continental drift prominent among them. Gutenberg had been an associate of Wegener's in Germany, before Gutenberg's move to Caltech in the late 1920s. However Gutenberg gave much less space to continental drift in his later book, *Physics of the Earth's Interior* published in 1959.

The only prominent advocates of continental drift in this period were Alex du Toit, in South Africa, and Sam Carey at the University of Tasmania. These two geologists enjoyed two advantages. One was that some of the clearest geological evidence for past continental connections exists in the southern continents, particularly the record of Permian glaciation showing the former existence of ice sheets immediately to the south. The other advantage was that, being located in the further reaches of the civilised Western world, they could perhaps be safely ignored in the important centres of learning. du Toit, especially, contributed a great deal of evidence and elaborated Wegener's ideas significantly in his book *Our Wandering Continents* in 1937, and his work attracted a significant minority of followers. On the other hand his tone was hectoring, according to Hallam, and this would have done little to endear his ideas to opinion leaders. One substantial modification of Wegener's scheme was that there had been two supercontinents rather than one: the northern Laurasia and the southern Gondwana. This reduced the presumed deformation of India and adjacent lands.

Carey also contributed important evidence and arguments, presented in a volume in 1958, *Continental Drift: a Symposium.* Carey was impressed by evidence for major compression and shear in New Guinca, where he did field work, and he was willing to be bold in contemplating major extensions and closures in and between other continents. Carey was so bold he proposed the radical idea that the Earth has expanded substantially since the Palaeozoic, this being his preferred mechanism for the evident pulling apart of continents that Wegener had already argued for. Carey's arguments were an important stimulus for Tuzo Wilson, and he was my introduction to tectonics before I ever imagined I would become involved myself, as I mentioned earlier.

Although continental drift was not entirely ignored after 1930, it was certainly very unfashionable and was dismissed by many geologists, often with some passion. Against this, it was widely recognised that a really satisfactory theory of mountain building did not exist. The old idea of a contracting Earth did not seem to provide for sufficient contraction to explain the observed crustal shortening, nor for zones of extension, without ad hoc elaborations of the theory. Expansion of the Earth was proposed by a few people, and occasionally mantle convection was appealed to in contexts other than continental drift. Most geologists worked on narrower problems, and little progress was made on the question of fundamental mechanism, despite much conjecture.

This situation prevailed until about the mid-nineteen-fifties, at which time two new kinds of evidence began to emerge that raised questions so serious they were harder to ignore. One kind of evidence was from paleomagnetism, the other from exploration of the sea floor.

When a rock forms, it can record the direction of the local magnetic field, because any grains of magnetic minerals incorporated into the rock tend to align with the field like a compass needle. Collectively these grains then produce a small magnetic field of their own that may be measurable in the laboratory. If a sufficiently large body of rock is magnetised in this way, the effect may even be measurable in the (geological) field as a detectable perturbation of the Earth's magnetic field.

Three distinct questions have been addressed through measurements of rock magnetism. First, have the rocks moved around on the Earth's surface? Second, has the magnetic field changed through time? Third, can rock magnetism be used to map the sequence of formation of rocks, or to date their formation? The second and third questions will be picked up later.

The first question was pursued by British geophysicists in the nineteen fifties, with a view to testing for continental drift. There were many complications to be dealt with, such as being sure that the original orientation of the rock could be reliably established and separating magnetisations acquired by the rock at different times through different microscopic mechanisms. There were also the possibilities that the magnetic field had not always been approximately aligned with the Earth's spin axis, that it had not always been approximately dipolar, as at present, and that the Earth had tilted relative to the spin axis.

By the late nineteen fifties, these difficulties had been substantially overcome. S. K. (Keith) Runcorn in 1959 reported strong evidence that North America and Europe had been closer together in the past, and Edward (Ted) Irving reported that Australia had moved nearly 90° northward from near the south pole. For those with knowledge of and confidence in the paleomagnetic data, this was strong evidence that continental drift had occurred. However, the earlier difficulties of the method were well-known, and it was hard for all but the minority involved in the measurements to know how much confidence to put in them. Nevertheless, these data, and many that followed, were very influential in reinstating continental drift as a respectable scientific topic.

Chapter 7
Like Nothing We've Seen Before

The second important source of evidence that revived the continental drift hypothesis came from exploration of the sea floor, which increased greatly during and after World War II.

Perhaps the most bizarre discovery was of magnetic 'stripes' on the sea floor. Ron Mason, of Imperial College, London and the Scripps Institute of Oceanography in California, was studying magnetism of oceanic sediments in the late 1950s. Because of this work, but still almost as an afterthought, a magnetometer was towed behind a ship doing a detailed bathymetric survey off the west coast of North America.

There emerged a striking and puzzling pattern of local variations in intensity of the Earth's magnetic field: over broad strips of the sea floor the magnetic field was alternately stronger and weaker. The variations are not large, and it was known on land that some rock formations are magnetised to a small degree, so their magnetism adds to the local field of the Earth. The pattern found by Mason was parallel to the local fabric of seafloor topography but was not directly expressed in topography. The magnetic variations were strong enough to reflect at least a kilometre thickness of basalt or similar rock. Mason presumed the strips of stronger magnetic field (black in Fig. 7.1) might reflect lavas deposited in elongated troughs, but their origin was obscure. The weaker (white) bands he presumed contained rocks with little magnetisation.

An intimate and insightful account of postwar exploration of the sea floor was given by H. W. (Bill) Menard, of the Scripps Institute of Oceanography, part of UC San Diego, in his 1986 book *The Ocean of Truth*. One of the early and most startling discoveries was the absence of thick sediment on the sea floor. If the continents and oceans were permanent features, there should have been a continuous sedimentary record of most of Earth history. Some geologists had eagerly anticipated an archive of the entire history of the planet, but few rocks older than the Mesozoic (200 million years) were found on the sea floor, and those have affinities suggesting they are fragments of continents.

Through the decade of the nineteen fifties, the global and continuous extent of the mid-ocean rise system was fully revealed: it is the second-largest surface feature on

G. F. Davies, *Stories from the Deep Earth*,
https://doi.org/10.1007/978-3-030-91359-5_7

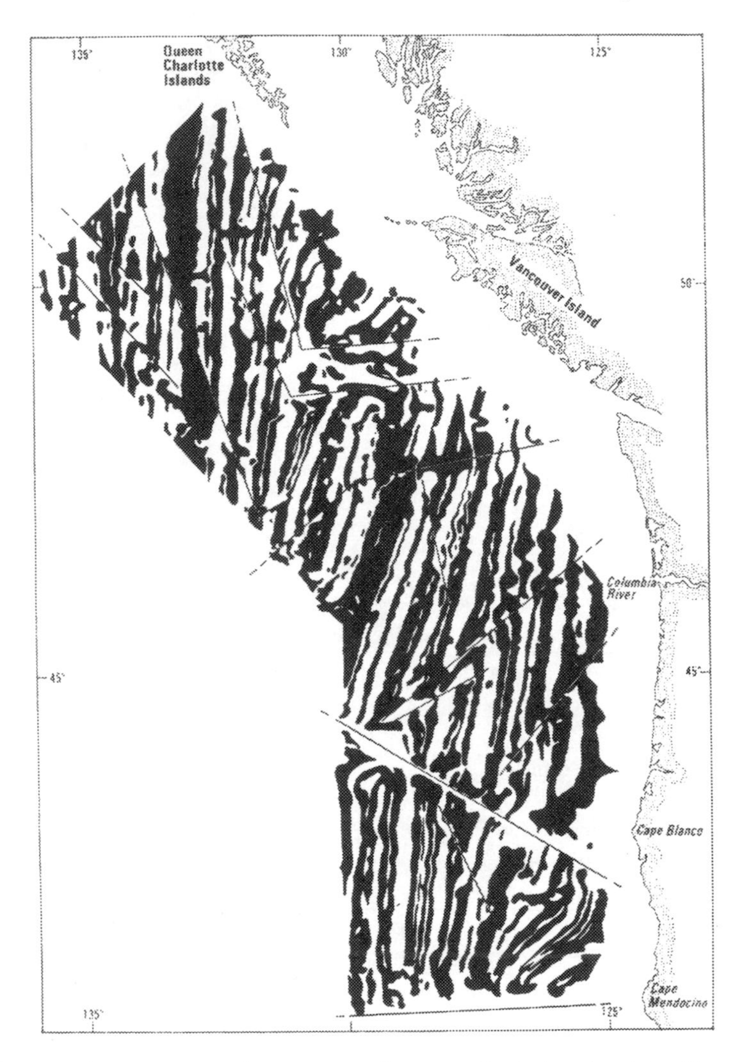

Fig. 7.1 Anomalies in magnetic field strength off western North America. After Mason [6] and Vine [7], reproduced with permission

the planet, after the continents. Great *fracture zones* were revealed on the sea floor. Fracture zones are narrow scars having the appearance of great faults thousands of kilometres long and remarkably straight. Vast areas of the sea floor, where it was not covered with thin sediment, comprised monotonously rough *abyssal hills* whose origin was unknown.

Guyots were found over a broad area of the central Pacific. Guyots are submarine mountains with flat tops. They were presumed to be of volcanic origin, and were and still are interpreted as former islands whose tops were eroded to sea level, after which they subsided below sea level. Harry Hess of Princeton University was another deeply involved in seafloor exploration, and both he and Menard inferred the former existence, about 100 million years ago, of a mid-ocean rise that has now subsided, taking the flat-topped islands down with it.

Menard called it the *Darwin Rise*, in honour of Charles Darwin. Darwin had observed coral islands of various types during his voyage on the *Beagle* in the 1830s. He inferred that they could be arranged into a time sequence, starting with an active volcanic island and progressing through an extinct island with a fringing coral reef, an eroded island with a barrier reef, and finally an *atoll*, comprising just a ring of reef, with the central island submerged out of sight. Darwin did not see such a sequence in a single chain, but J. D. Dana did observe such sequences, and Darwin's interpretation became the accepted account. This is rated by some as Darwin's most important contribution to geology, as distinct from his theory of natural selection of organisms.

The heat flux conducted through the sea floor was found to be as large or larger than on continents, despite the continental crust having a considerably higher content of radioactive heat sources.

Many of these discoveries were quite unexpected and difficult to make sense of. We must realise that the area being explored was vast, and that the picture was at first patchy and incomplete. Nevertheless it was clear that old ideas had to be revised in major ways. The sea floor does not look much like the continents geologists were used to. Any typical piece of the continental crust has been repeatedly contorted, faulted, buried and heated, intruded with magmas, exhumed and eroded, perhaps with a relatively thin covering of relatively undeformed sediments. Long linear structures and broad areas of similarity are rare. In comparison the sea floor was turning out to be positively orderly, but not in any way that made sense. Figure 7.2 is a map of the northeast Pacific sea floor showing major fracture zones, along with magnetic anomalies that we will return to a little later.

The fracture zones are uniquely long and linear features, and it is hard to interpreted them as anything other than strike-slip faults with large displacements across them, as revealed by magnetic anomalies in Fig. 7.2. Some of the Pacific fracture zones extend for around 5000 km towards the Hawaiian Islands, and magnetic anomalies are offset by hundreds of kilometres in some places. Long strike-slip faults (in which blocks of crust slide horizontally past each other) are known on land, such as the San Andreas fault in California or the Anatolian fault in Turkey, but they extend for no more than about 1000 km. Even more strange, the fracture zones seem to disappear at continental margins and to have no obvious extension into the continents, despite the appearance of very large faulted offsets across them.

The thin sediment covering on the sea floor required either that the rate of sedimentation had been very much less in the past than at present or that the sea floor is no more than about 200 million years old. The abyssal hills topography looks chaotic, suggesting widespread tectonic disruption, at least on a small scale, but the sediments overlying them on the older sea floor are largely flat-lying and undisturbed.

The relationship of fracture zones to mid-ocean rises, if any, was unclear until later. In the north-east Pacific several major fracture zones connect to nothing obvious at either end. In the east they run up to the edge of the continent and appear to stop, while

Fig. 7.2 Fracture zones and magnetic anomalies of the northeast Pacific sea floor, including the area covered in Fig. 7.1. Fracture zones are the heavy lines, mostly running east–west. The lighter lines are magnetic anomalies, identified by their number in a long sequence according to the key on the right. Magnetic anomalies (e.g. number 20) can be seen to be offset across fracture zones, sometimes by hundreds of kilometres. There are complications near the continental margin, and many detailed features, some of whose interpretation will be discussed later. From Atwater and Severinghaus [8], reproduced with permission

in the west they peter out. In the Atlantic, relatively rough seafloor topography and mostly east–west surveys left the picture confused, with Bruce Heezen of Columbia University inferring that east–west troughs were part of a continuous graben on the rise crest. Only later were they interpreted as fracture zones offsetting the rise crest.

According to Marcia Bjornerud the first to recognise the extensional nature of the rise crests was Marie Tharp. She was the person who, in the days before automated everything, converted numbers from bathymetric sounding profiles into meticulous topographic maps of the sea floor. She had a degree in geology and worked for many years at Columbia University turning numbers into maps. Her suggestion that the common trough at the crests of the rises was a graben, a down-dropped block caused by extension, was at first dismissed by Bruce Heezen as 'girl talk'. Tharp was a coauthor on papers, but it was Heezen who got most of the credit for interpretation.

The origin of the mid-ocean rise system was obscure. Where it was traced onto land in Iceland and East Africa, it was undergoing extension. This was consistent with the presence of an axial trough along much of the crest of the mid-Atlantic rise, and Heezen in 1960 inferred that the entire system of rises was extensional. However for some time seismic reflection data seemed to show a covering of sediment over the East Pacific Rise, and Menard inferred that it might be young and had not yet begun active rifting. Menard and Hess inferred that rises are ephemeral, and Menard proposed that the East Pacific Rise is young, that the Mid-Altlantic Rise is mature, with active rifting, and the Darwin Rise is extinct.

Menard and Hess proposed variations on ephemeral convective upwellings of the mantle to explain the existence of the rises. Heezen had traced the Mid-Atlantic Rise around Africa and into the Indian Ocean and had inferred that it is all extensional. He reasoned from this that the Earth had to be expanding, otherwise Africa would be undergoing active compression because of being squeezed from both sides by the extending rises.

The idea that the sea floor is or has been mobile was implicit in the interpretation of rises and fracture zones. The uniformity of the sea floor and the absence of widespread evidence of deformation of sediments suggested that large areas of it were moving coherently. For example, Menard thought that the long slices between fracture zones moved independently, driven by separate convection 'cells'.

This brief summary, following Menard's account, gives some flavour of the ferment of ideas that was induced by the new kinds of observations. They were so puzzling, especially while they were incomplete, and sometimes misleading, that people were willing to appeal even to such disreputable ideas as mantle convection or Earth expansion. Menard points out, however, that most other geologists at the time were busy with other things and unaware of or unconcerned with the sea floor, and that the oceanographers' research also was 'narrow, mostly marine geomorphology, but the areas were hemispheric and the conclusions correspondingly grand'.

That is the context in which two people proposed a third type of explanation for the mid-ocean rises. Not Earth expansion and not ephemeral convection cells, but continuous convection, coming right to the Earth's surface at rise crests and descending again at deep sea trenches. Harry Hess wrote his paper in 1960, but it was not published until 1962, while Robert Dietz' paper was written and published in 1961. Menard argues persuasively that their work was independent.

Hess and Dietz accepted Heezen's arguments that the mid-ocean rises are extensional rifts, but they did not accept his conclusion that the Earth expands. Hess had a long-standing interest in deep ocean trenches. Vening Meinesz had measured gravity at sea in submarines, and found large negative gravity anomalies over trenches that he attributed to a down-buckling of the crust where two mantle convection currents converged. He developed a 'tectogene' theory that trenches were the early stages of *geosynclines*, where thick sediments accumulated, later to be thrust upward in association with volcanic activity. Dietz also had an interest in geosynclines, arguing in later papers that they represent former passive continental margins that are activated by subduction. Thus both Hess and Dietz were disposed to the idea of crustal convergence and descending convection at trenches.

The central ideas that have survived from these papers are that convective upwelling of the mantle reaches the surface in a narrow rift at the crest of mid-ocean rises and forms new sea floor. This then drifts away on both sides of the rift, ultimately to descend again into the Earth at an ocean trench. A continent can be carried passively by the horizontal part of the convection flow, rather than having to plow through the sea floor, as supposed by Wegener. The youth of the sea floor and the thinness of sediments would thus be accounted for. A uniformly thick crust might be formed, if it is all formed by the same process at a rise crest. Dietz recognised that the abyssal hills topography might also be a residue of rifting at the rise. The high heat flow on rises would be explained by the close approach of hot mantle to the surface. Dietz coined the concise term *seafloor spreading*.

How did these radical ideas get published, and gain traction, when 'mobilist' ideas in general were widely held to be disreputable? Hess was evidently conscious of the potential reaction, because in his introduction he said 'I shall consider this paper an essay in geopoetry'. He also vowed to hold as closely as possible to a uniformitarian approach. The first claim was false modesty and both were clearly intended to appease potential attackers—'Oh don't take me too seriously, and please don't lump me with the catastrophists'.

Hess was a quite distinguished professor at a premier US university, Princeton, so in fact his word carried some weight. He was also synthesising a lot of new observations of the sea floor where good information had been sparse. The paper was published in a special volume honouring a retiring colleague, so would have been less rigorously refereed. So he had less reason to worry than most, but his interpretations and hypotheses were quite radical. The fact that Dietz published the idea of seafloor spreading at the same time presumably helped them both. Dietz' paper, in contrast to Hess', was in the high-visibility British journal *Nature* and was more businesslike in its presentation. *Nature* was more inclined to publish stimulating ideas, and mobilism was not quite as unpopular in Britain, thanks in part to Arthur Holmes' widely-used text book, which we will get to shortly. The puzzling findings from seafloor exploration had also been attracting some wider attention, so for some combination of these reasons the idea of seafloor spreading was not dismissed out of hand.

Not all of the ideas from Hess' paper have survived. For example, the composition of the oceanic crust was not definitely known at the time, and he supposed it to be serpentine (hydrated mantle peridotite), whereas Dietz more correctly assumed it to be basalt produced by melting the mantle under the rise. Hess still thought rises were ephemeral, being misled by the assumption that the Darwin Rise was of the same type as the modern mid-ocean rises. The Darwin Rise loomed large in Hess' thinking because of his discovery of guyots. Some have viewed guyots as a key link to the idea of seafloor spreading (for example Alan Cox in his 1973 anthology *Plate Tectonics and Geomagnetic Reversals*), but I think they distracted him into thinking more about vertical motions than horizontal, and his thinking was still a bit confused in this paper. Hess did not think fracture zones were related to rises. Dietz did, and he proposed that the convection proceeded at different rates on either side of a fracture zone, so the sea floor is displaced by different amounts; Menard had a similar conception.

Hess made another fundamental, though somewhat separate point in his paper: that continents would tend to be piled up by convection and then eroded down towards sea level. The consequence would be that the level of the continents would be near sea level, and this would be the result of a dynamic equilibrium between the piling up and the erosion. Thus he correctly recognised the explanation for the bimodal distribution of the elevation of the Earth's surface that had been an important argument of Wegener's.

It may seem curious that Hess' and Dietz' papers became famous for proposing seafloor spreading, but not for the complementary removal of sea floor at trenches, which was an integral part of their concept. The reason is probably that the understanding of trenches and their associated mountains (island arcs or active continental margins) was in a state of confusion at the time, and as a result neither of them put much stress on what we now call subduction. Although there was a widespread concept that trenches were the sites of compression and some downward buckling, according to Vening Meinesz, or faulting, according to Benioff, the amounts of crustal motion they envisaged was usually limited. As well, attempts to determine the direction of slip in earthquakes from seismic waves were yielding confusing and inconsistent results. It was not until after a world-wide network of standardised seismographs was in place in about 1963 (to monitor underground nuclear explosions) that clear results of this type emerged. However the confusion did not hinder Wilson, as we will see later.

This account of seafloor spreading has been expressed very much in terms of mantle convection, because that is how Hess and Dietz conceived it. You will see in the next chapter that there are advantages in looking just at the surface of the Earth, without worrying about what is happening underneath. However, the question of how mantle convection relates to the surface became more acute as the surface picture was clarified. Already with seafloor spreading there is the novel idea that mantle convection rises right to the Earth's surface, but only in a very narrow rift zone at the crest of a mid-ocean rise. This is a novel form of convection. Holmes' picture was rather similar to that of Hess and Dietz, even to the point of having a regenerating basaltic oceanic crust, but his concept was conditioned by conventional ideas about convection, and he supposed that seafloor extension occurred over a broad region. Holmes' sketch is unclear about what happens where convection turns downward, but he had some penetrating thoughts about that, as we will see.

The idea of convection in the Earth's mantle goes well back into the nineteenth century. An early mention of mantle convection is by W. Hopkins in 1839. Reverend Osmond Fisher, in his 1881 book *Physics of the Earth's Crust*, proposed mantle convection as a tectonic agent, with flow rising under the oceans and descending under continents. He assumed the mantle to be relatively fluid, drawing on the concepts of isostasy being developed at that time. He envisaged that this flow would expand the oceans and compress the continents at their edges, generating mountains.

According to Hallam in *Great Geological Controversies* (1989), the idea of a fluid mantle was more widespread in continental Europe, particularly in Germany, than in Britain and America. He cites a number of instances of this, noting that this implies a more sympathetic climate around the turn of the century within which Wegener's ideas of continental drift could develop. However, Wegener himself did not appeal to mantle convection, and his concept that continents plough through oceanic crust seems to owe little to any idea of a deformable mantle.

It was Arthur Holmes who most seriously advocated mantle convection, and he proposed it explicitly as a mechanism for continental drift, first in a talk and brief note in 1928, then in a paper in 1931, and finally in his book *Principles of Physical Geology*, the first edition of which appeared in 1944. It is instructive to see how creative was Holmes' thinking.

Holmes' basic proposal was that convection occurs under the lithosphere and drags the continents around. His proposed flow was different from Fisher's, in that Holmes, in his initial version, reasoned that convection might rise under a continent because of the thermal blanketing effect of continental radioactivity, a subject that he was very familiar with, having pioneered radiometric dating.

Holmes then envisaged that the rising and diverging convection might rift a continent and carry the pieces apart. In his earlier version, he supposed that a piece of continent might be left over the upwelling site, because the horizontal flow would be relatively stagnant there. In his later version, he proposed instead that the crust between the diverging continental fragments might be broadly stretched and the extension accommodated by the intrusion or eruption of basaltic melts generated in the (presumed) warmer upwelling mantle.

Holmes also envisaged that a basaltic oceanic crust would be returned to the mantle. He presented the case with admirable simplicity:

> The obstruction that stands in the way of continental advance is the basaltic layer, and obviously for advance to be possible the basaltic rocks must be continuously moved out of the way. In other words, they must founder into the depths, since there can be nowhere else for them to go.

In this later version Holmes proposed a different driving force for his convecting system. He contrasted sialic (continental) rocks, whose density was not known at that time to be much affected by pressure, with basaltic compositions, which were known to be converted by pressure first to granulites and then to eclogite, undergoing in the process a density increase from about 2.9 to 3.4 Mg/m^3 (mega-grams per cubic metre, the same as grams per cubic centimetre). It was not known at that time that the oceanic basaltic crust is quite thin (about 7 km), so his proposal was quite plausible:

> Since this change is known to have happened to certain masses of basaltic rocks that have been involved in the stresses of mountain building, it may be safely inferred that basaltic roots would undergo a similar metamorphism into eclogite. Such roots could not, of course, exert any [positive] buoyancy, and for this reason it is impossible that tectonic mountains could ever arise from the ocean floor. On the contrary, a heavy root formed of eclogite would continue to develop downwards until it merged into and became part of the descending current, so gradually sinking out of the way, and providing room for the crust on either side to be drawn inwards by the horizontal currents beneath them.

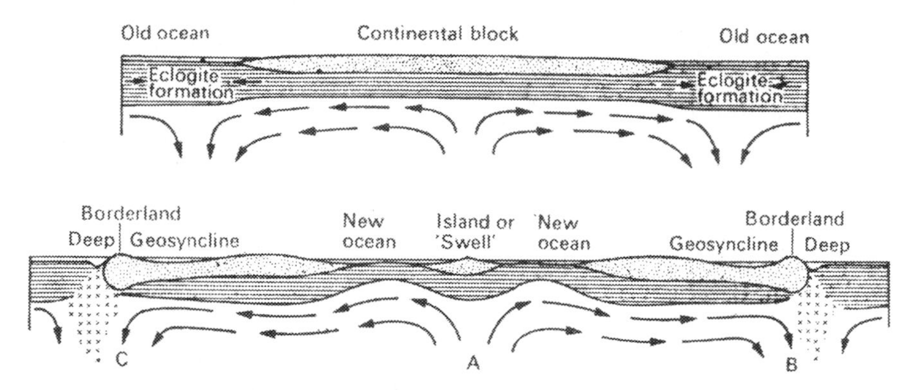

Fig. 7.3 Arthur Holmes' conception of how mantle convection might rift a continent and carry the pieces along, forming new sea floor between and building mountains (at the borderland and geosyncline). From Holmes [9]

Thus Holmes, in this later version, proposed the generation of a basaltic crust over mantle upwellings and its removal into downwellings, concluding

> To sum up: during large-scale convective circulation the basaltic layer becomes a kind of endless travelling belt on the top of which a continent can be carried along, until it comes to rest (relative to the belt) when its advancing front reaches the place where the belt turns downwards and disappears into the Earth.

Menard commented on how closely this anticipates Dietz' version of seafloor spreading, the only essential difference being that Dietz proposed that the basaltic oceanic crust is produced in the narrow rift zone at the crest of the mid-ocean rise system, whereas Holmes assumed it would emerge over a broad extensional area. Dietz of course had far more complete observations of the sea floor available, including the evidence for rifting of the crest of the mid-ocean rises.

Holmes' ideas were not entirely ignored, although they did not become part of mainstream thinking. In 1935 evidently some young turks held a symposium in London on continental drift, some of which was published by the Royal Society (in a Monthly Notices Supplement). C. L. Pekeris showed that convection driven by the differential thermal blanketing of continents and oceans could result in velocities of millimetres per year and stresses sufficient to maintain observed long-wavelength gravity anomalies. Anton L. Hales showed that plausible convection could be maintained by a mean vertical temperature gradient (above the adiabatic gradient) of as little as 0.1 °C/km. In North America N. A. Haskell's 1937 estimate of mantle viscosity from post-glacial rebound appeared during this period, while David Griggs in 1939 developed a number of ideas, a central one being that experimentally observed non-linearities in rock rheology could result in episodic convection. He also presented a simple laboratory realisation of the way crust might be piled into mountains over a convective downwelling, analogous to Holmes' conception (Fig. 7.3). This experiment may have had a positive influence on concepts of subduction and the interpretation of the Wadati-Benioff deep earthquake zones.

By the mid-1960s the idea of continental drift was being seriously revived, and continents were being seen as rafted along by convection in the mantle. Upwelling at mid-ocean rises and descent at deep ocean trenches were explicitly invoked. Yet the picture of the surface motions was about to be clarified in a startling way, but in a way that made the relationship with convection only more puzzling.

Chapter 8
Novel Ideas: Plates and Plumes

J. Tuzo Wilson was a Canadian physicist turned geologist. He is well known for recognising a new class of faults, and for naming them 'transform faults', in a paper published in *Nature* 1965. The paper is widely recognised as providing a key link in understanding tectonics. However it is more than that, because in this paper the concept of plate tectonics first appears in its complete form. Not only was the system of plates sketched out, but their form was quite novel, and even somewhat strange as details were clarified.

The paper is titled *A new class of faults and their bearing on continental drift*. It is worth quoting the opening of the paper.

> Many geologists have maintained that movements of the Earth's crust are concentrated in mobile belts, which may take the form of mountains, mid-ocean ridges or major faults with large horizontal movements. These features and the seismic activity along them often appear to end abruptly, which is puzzling. The problem has been difficult to investigate because most terminations lie in ocean basins.
>
> This article suggests that these features are not isolated, that few come to dead ends, but that they are connected into a continuous network of mobile belts about the Earth which divide the surface into several large rigid plates.

Here, Wilson has defined the problem and presented its solution with simple clarity. His sketch map (Fig. 8.1) gave the world its first view of the tectonic plates.

Wilson had very broad interests in geology, but in particular he had been investigating large transcurrent faults, which are long faults with mainly horizontal relative motion across them—so they are large strike-slip faults to which geologists found it convenient to give a special name. He actually had not believed in continental drift until about 1960, but the publication of Dietz' seafloor spreading paper in 1961 convinced him that it must be right. He set about finding more evidence, from ocean islands and from transcurrent faults, as he recounted in a 1963 article in *Scientific American.*

He recognised that if the North Atlantic had been closed, as continental drift advocates proposed, then the Cabot fault in eastern North America would match up with the Great Glen fault that cuts across the top of Scotland. He was puzzled, like

G. F. Davies, *Stories from the Deep Earth*,
https://doi.org/10.1007/978-3-030-91359-5_8

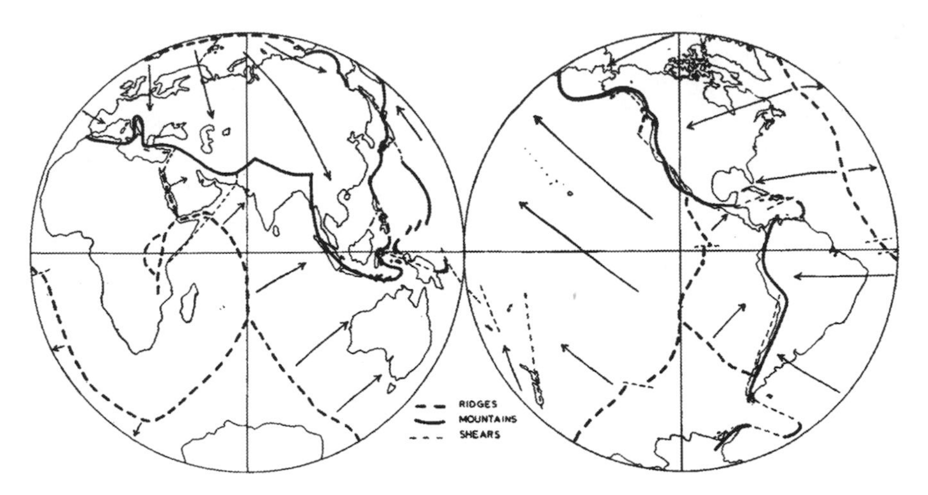

Fig. 8.1 Wilson's sketch map from 1965. The original caption is as follows. 'Sketch map illustrating the present network of mobile belts around the globe. Such belts comprise the active primary mountains and island arcs in compression (solid lines), active transform faults in horizontal shear (light dashed lines), and active mid-ocean ridges in tension (heavy dashed lines). From Wilson [10], reproduced with permission

others, by the great fracture zones that were being discovered on the ocean floor, because they seemed to be transcurrent faults of large offset, but they stopped at the continental margin, with no equivalent expression on the adjacent continent.

Wilson's clinching insight was his recognition of the way these great faults can connect consistently with mid-ocean ridges or with 'mountains' (meaning subduction zones) if pieces of the crust are moving relative to each other as rigid blocks, without having to conserve crust locally; in other words the blocks might grow or shrink. Continuing the above quotation,

> Any feature at its apparent termination may be transformed into another feature of one of the other two types. For example, a fault may be transformed into a mid-ocean ridge as illustrated in [Fig. 8.2]. At the point of transformation the horizontal shear motion along the fault ends abruptly by being changed into an expanding tensional motion across the ridge or rift with a change in seismicity.

'… with a change in seismicity'? I'll return to that (Fig. 8.2).

Wilson explains how his 'transform' faults may connect a ridge to a subduction zone (*trench*, for short), or to another ridge segment, or may connect two trenches. He points out that the sense of motion on a transform fault joining two ridge segments is the reverse of the superficial appearance (as we saw in Chap. 3, Fig. 3.3), and that the traces left by such faults beyond the ridge segments they connect are inactive. Transform faults may grow or shrink in length as a simple consequence of symmetric ridge spreading and asymmetric subduction. He does a fast tour of the world, explaining relationships between major structures, elucidating what we now know as plate boundaries.

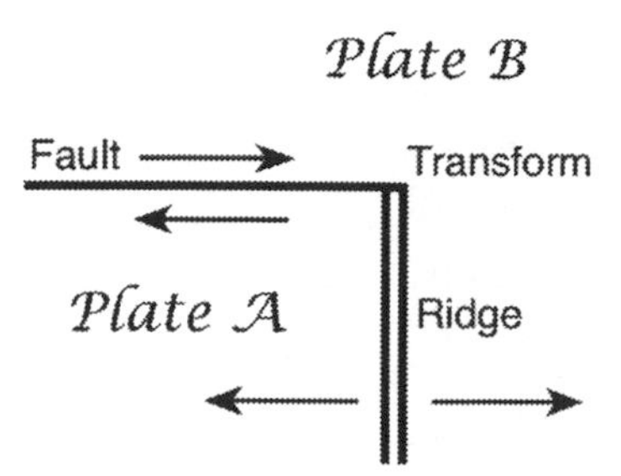

Fig. 8.2 Illustration of Wilson's idea of how a transform fault (Fault) is transformed at a point (Transform) into a mid-ocean ridge spreading centre (Ridge). The concept depends on the pieces of crust on either side of F and R (*Plate A* and *Plate B*) moving independently as rigid blocks or plates without requiring the area of each plate to be conserved locally

The language of the paper is terse. One senses the excitement of the rush of insights as pieces of a puzzle (literally) fall into place, and the desire to pack as much as possible into a short, crucial paper. Key information is almost lost. He forgets to spell out that it was known that the only seismically active parts of the great fracture zones cutting across the equatorial Atlantic sea floor are the parts between the ridge segments, as we explored in Chap. 3, and that this was a major puzzle. That information appears only in the caption of a sketch map, and ambiguously, where he distinguishes active faults as solid lines and 'inactive traces' as dashed lines, without making clear that this had already been observed, and was not just a prediction of his theory. The cryptic 'with a change in seismicity' noted above means that the type of earthquake changes from strike-slip to *normal faulting* (extensional) where a transform fault joins a ridge segment.

A crucial difference between Wilson's thinking and what had gone before was that he was thinking as a structural geologist, looking only at the surface, and not worrying about what might be happening underneath. He envisaged rigid blocks bounded by three types of boundary that correspond to the three standard geological fault types: *strike-slip* (transform fault), *normal* (ridge) and *reverse* (subduction zone).

Conceptually he narrowed the older notion of *mobile belts* down to sharp boundaries. He explicitly adopted the long-standing implication of that term, that there is little deformation outside the mobile belts, taking it conceptually to the limit of proposing that there is no deformation. He was explicit in the fourth sentence, quoted above, that the plates are 'rigid'. The point is explicit also within the paper: 'These proposals owe much to the ideas of S. W. Carey, but differ in that I suggest that the plates between the mobile belts are not readily deformed except at their edges.'

In retrospect it seems like an obvious inference of the recognition of mobile belts, that the regions between do not deform. Geologists certainly had the concept of a non-deforming *shield*, but applied it to regions that had not deformed for a geologically long period (a billion years or more), such as the Canadian shield, rather than to present inactivity. It is true that the sea floor had been only recently explored, but Gutenberg and Richter had published a global map of earthquake occurrence in 1941, updated in 1954. Their map clearly shows earthquakes mostly confined to relatively

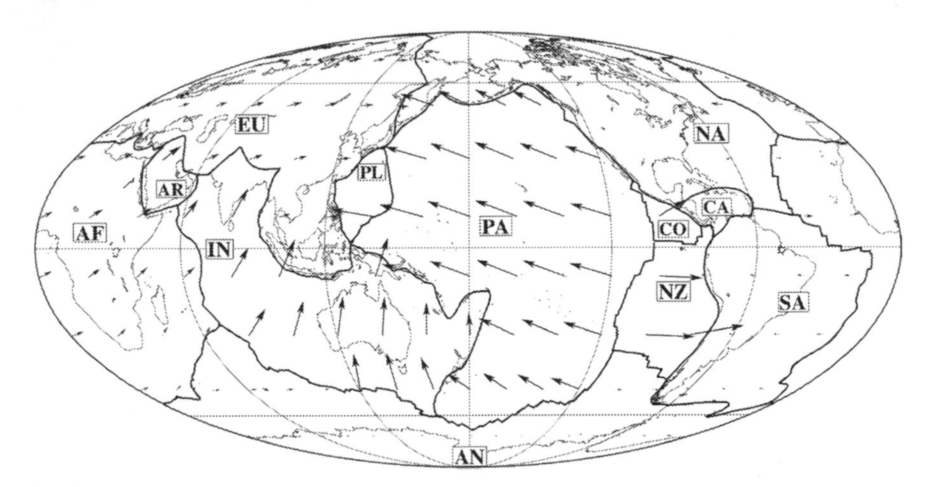

Fig. 8.3 A map of the plates and their velocities from 1998. The irregular, angular shapes and the range of sizes do not look like the simple, early conceptions of convection of, for example, Holmes in Fig. 7.3. Note especially the banana-shaped India-Australia plate. From Lithgow-Bertelloni and Richards [11], reproduced with permission

narrow belts, with little activity in between. Indeed their map outlines most of what Wilson identified as a continuously connected network of plate boundaries.

Wilson's focus on the surface was why he was able to see the plates in all their simplicity. By focussing on the surface motions he recognised the behaviour of a brittle solid, and successfully defined plate tectonics in those terms. However this clarity at the surface came at the cost of understanding how it relates to the underlying mantle. Whereas Holmes, Hess and Dietz conceived the surface motions as expressions of convection of a fluid mantle, the plates look nothing like fluid convection. Figure 8.3 shows a more complete and precise map of the plates from 1998.

In Wilson's conception we had a two-dimensional view of the Earth, namely its spherical surface. In Holmes' and others' cartoons of mantle convection (Fig. 7.3) we had two-dimensional cross-sections. How did they go together to make a coherent three-dimensional picture? This became a central puzzle for the next couple of decades, usually posed as 'What is the driving mechanism?' It was widely presumed that mantle convection was involved, but there was great confusion about how exactly it related to the plates.

Wilson's scheme is not about a driving mechanism. It is a description of the motions of plates. It can be called a *kinematic* theory, a theory of motions. A theory of how the plates are driven would be called a *dynamic* theory, a theory involving forces. Kepler's theory of the planets moving in elliptical paths around the sun is a kinematic theory, whereas Newton's description invokes gravitational forces (at a distance) and it is a dynamical theory.

To call Wilson's conception a kinematic theory is not be belittle it. It was a revolutionary view of the motions of the crust, and Wilson was clearly aware of the implications, justifying the second part of his title ' ... and their bearing on continental drift'. That is explicit in his explanation of the transform concept and in the last sentence of the paper where, referring to transform faults, he says 'proof of their existence would go far towards establishing the reality of continental drift and showing the nature of the displacements involved.' Perhaps too modestly, he implies here that he has not already pointed out compelling evidence, in the form distribution of earthquakes on fracture zones and implicitly in the wealth of geological and seismological evidence that had given rise to the concept of mobile belts and the complementary idea of internally stable blocks.

Contrast Wilson's paper with a little-known paper by fellow-Canadian A. M. Coode, also published in 1965. In this very brief note, Coode elegantly presents the conception of a ridge-ridge transform fault, along with a diagram explaining how both the ridge crest and magnetic anomalies (discussed later) are offset. That is all Coode does. The further implications are not developed. The paper was almost unknown until it was pointed out by Menard in his book, though this was also because it was in a journal where oceanographers were unlikely to see it. It was an important contribution, but overlooked.

There can be no doubt also that Wilson appreciated that he had taken a major step towards a unifying dynamic theory of the Earth that would probably involve mantle convection. Two years earlier he had published several papers containing the fruits of a remarkable burst of creativity, including the seminal insight that led to the idea of mantle plumes, and a wide-ranging article in *Scientific American* on continental drift. In the latter it is clear that he has a comprehensive grasp not only of a large number of geological observations but also of the arguments from isostasy, post-glacial rebound, materials science and gravity observations over ocean trenches that the mantle is deformable and undergoing convection. His map of convection currents bears a strong resemblance to his 1965 map of the plates, and he writes of moving crustal blocks.

Reading the 1965 paper, we may see a structural geologist presenting a brilliant and novel synthesis. Reading it in conjunction with the 1963 papers, we see more: a scientist in the full pursuit of the secrets of the Earth, chasing whatever kind of evidence will serve. Reading them all, I see a man move, in little more than five years, from first conversion to mobilism through to clarity of understanding of geology's major unifying concept. I think he deserves a special place in the pantheon of geology for being the first to see the plates in complete and simple form.

Comparing the plate-tectonic revolution to the Copernican revolution in his preface to a 1976 collection of *Scientific American* articles, Wilson made the following observation.

> That the earth is the centre of the universe and that it rests on a fixed support was the obvious and early interpretation. To realise that the earth is spinning freely in space and that the sun, and not the earth, is at the focus of the solar system required a prodigious feat of imagination. ... Changing the basic point of view created a new form of science with a different frame of reference. It was this change in the manner of interpreting the observations that constituted the scientific revolution.

Wilson was the first to complete the change in point of view to a coherent picture of crustal motions, one that was immediately perceived as fruitful, making sense

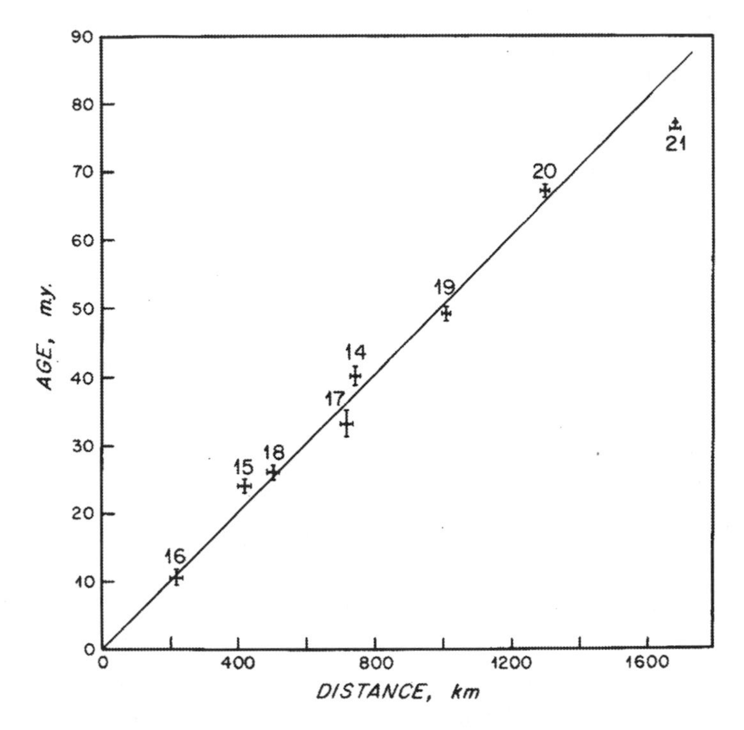

Fig. 8.4 Age of the lowermost sediment *versus* distance from the Mid-Atlantic Ridge. From a 1970 profile by Maxwell and others [12], reproduced with permission

of a wealth of accumulated observations. The paper by Isacks, Oliver and Sykes discussed in Chap. 3 is one example. It did of course build on Wegener's conception of continental drift, itself built on a diversity of telling observations, notwithstanding their relative sparsity at the time and the imperfections in details of his conception. Hallam's 1973 book title captures their joint accomplishment: *A Revolution in the Earth Sciences: From Continental Drift to Plate Tectonics.*

Three kinds of observational confirmation of seafloor spreading, and implicitly of plate tectonics, accumulated through the 1960s. The easiest to appreciate, and the evidence that persuaded many geologists unfamiliar with geophysical evidence, was the determination of sediment ages across the South Atlantic sea floor, presented by A. E. Maxwell and others in 1970. The sediments were drilled at a series of locations. The lowermost sediment was presumed to have been laid down soon after the local sea floor was formed at the Ridge. Ages were determined from small marine fossils. The resulting plot, Fig. 8.4, shows a clear and consistent trend to greater age with distance from the Ridge, exactly as predicted by seafloor spreading. This was the kind of definitive test that Wilson had sought through the ages of oceanic islands.

We have already touched on a second line of evidence, from the earthquakes along mid-ocean rises. The work summarised by Isacks, Oliver and Sykes (Fig. 3.1) showed two key things. First, earthquakes are confined to the *spreading centre* crests of rises and to the portions of fracture zones connecting those crests, with very few on fracture zones where they extend beyond spreading centres. Second, the earthquakes on the active portions of fracture zones are of the sense predicted by Wilson's transform fault interpretation.

To elaborate the latter point, it became possible in the early 1960s to obtain reliable records of seismic waves from around the world, through the World Wide Network of Standard Seismographs set up to monitor underground nuclear explosions. In turn this enable the *radiation pattern* of an earthquake to be determined: in some directions the first wave is compressional, in other directions it is extensional. (The radiation refers to radiated seismic waves, not to anything more sinister.) The radiation pattern can distinguish left-lateral fault slippage from right-lateral. So for example in Fig. 3.3 we expect left-lateral fault motion, rather than the right-lateral offset suggested by the ridge crests. Earthquakes analysed by Lynn Sykes over several years strongly confirmed the transform fault interpretation, which in turn confirmed the sea-floor spreading interpretation.

The third line of evidence concerns those strange magnetic stripes on the seafloor (Figs. 7.1 and 7.2), and this requires a little more explanation. Two groups in the early 1960s were investigating the magnetisation of basaltic lava sequences, one group at the US Geological Survey in California and the other group at the Australian National University. They were following up a report by M. Matuyama in 1929 that the uppermost lavas of a sequence were magnetised parallel to the local direction of the Earth's magnetic field, but lower and older lava flows were magnetised in the opposite direction. Matuyama suggested that the Earth's magnetic field had been reversed in the past, so magnetic 'north' was at geographic south, but this claim had been vigorously disputed. With the aid of the newly established potassium-argon dating method and the latest magnetic techniques, the later groups were able to confirm Matuyama's observation, and the interpretation that indeed the magnetic field of the Earth had been reversed about a million years ago. They went on to establish a sequence of reversals over the past 4–5 million years, about one reversal per million years.

Meanwhile Fred Vine, a graduate student at Cambridge University, was analysing data from a magnetic survey over the Carlsberg Ridge in the Indian Ocean, collected by his supervisor Drummond Matthews. He noticed that a seamount near the ridge crest was reversely magnetised. This is easier to infer for seamounts, because they are more like point sources and produce a more distinctive three-dimensional pattern of anomalies, whereas a long strip of sea floor produces a two-dimensional pattern that is more ambiguous. The survey had also revealed magnetic stripes, of the kinds seen by Mason, parallel to the ridge crest. There was a stripe over the ridge crest, then a gap on either side, then another stripe, in a roughly symmetrical pattern. While his supervisor was away, and with the reversely magnetised seamount as a clue, Vine conceived an explanation for the magnetic stripes that is illustrated in Fig. 8.5.

If seafloor spreading is occurring at the crest of the Ridge, then new basaltic crust is being formed and will be magnetised parallel to the local direction of the Earth's magnetic field (Fig. 8.5a). If, later, the Earth's field reverses, then new sea floor will form in the middle of the sea floor just formed, but will be reversely magnetised.

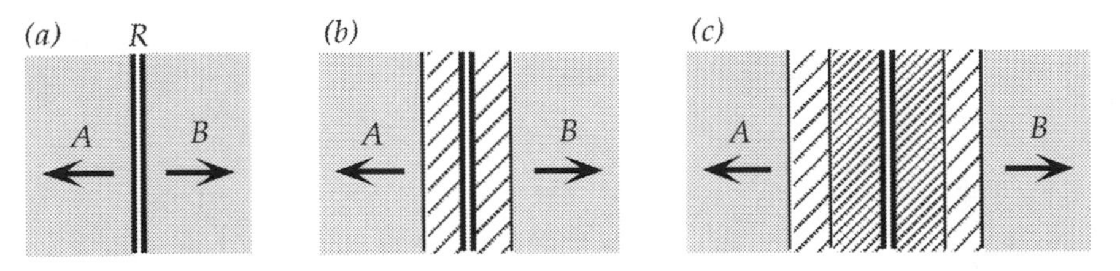

Fig. 8.5 Illustration in map view of the way sea floor spreading and magnetic field reversals combine to yield strips of sea floor that are alternately magnetised in normal and reverse directions. The resulting pattern is symmetric about the crest of the ridge if the spreading itself is symmetric (meaning that equal amounts of new sea floor are added to each plate). From Davies [4], reproduced with permission

This yields a *reversely* magnetised stripe that splits the original *normally* magnetised stripe (Fig. 8.5b). If the magnetic field returns to its 'normal' direction, then in turn the second stripe will be split by a normal stripe (Fig. 8.5c). The result is a pattern of alternating normal and reverse stripes, and the pattern is symmetric about the ridge crest. This interpretation was published in 1963 in a brief and famous paper in *Nature*. It turned out that Lawrence Morley in Canada had the same idea, but was unable to get it published: in fact it was rather rudely rejected by at least one reviewer.

Vine's prediction of a symmetric pattern of stripes was confirmed spectacularly by a survey south of Iceland published by James Heirtzler and others in 1966 (Fig. 8.6). Vine later commented that his interpretation required three assumptions each of which was, at the time, highly controversial: seafloor spreading, magnetic field reversals, and that the oceanic crust was basalt and not consolidated sediment.

Vine went on to analyse other surveys, including Ron Mason's from the Pacific (Fig. 7.1). He was able to find two small active spreading centres near the continental edge, and small symmetric sections of the pattern. The rest of the pattern, it was implied, had formed at a spreading centre that no longer existed, having been over-ridden by the North American continent as it drifted westward. The further implication is that there used to be a large plate subducting under North America: what is now known as the Farallon plate. The broader patterns in Fig. 7.2 record, among other things, how a spreading centre between the Pacific and Farallon plates had previously existed for millions of years.

Subsequent exploration revealed extensive patterns of magnetic stripes in all the main ocean basins, and that correlated with the chronology of magnetic field reversals established on land. These magnetic stripes allowed ages to be assigned to vast areas of the sea floor on the basis of the reversal sequence.

We can reflect on the magnitude of that last paragraph. Assigning ages to rocks always has been and still is a central occupation of geologists. It is painstaking work, whether the method is correlation of fossils or measurement of radioactive decay. It took most of the twentieth century to develop the ability to get reliable ages accurate to within a few percent or less for many kinds of rocks. As Menard remarked

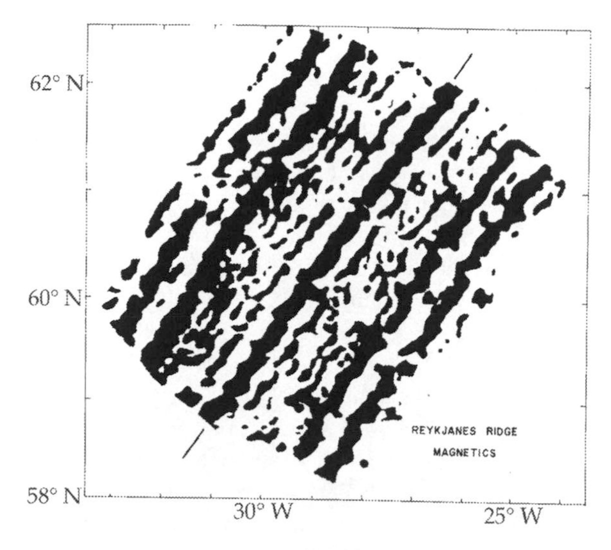

Fig. 8.6 The pattern of magnetic anomalies across the Mid-Atlantic Ridge south of Iceland, where it is known as the Reykjanes Ridge. Black indicates a positive anomaly, inferred to be due to normally magnetised crust, and white indicates a negative anomaly, inferred to be due to reversely magnetised crust. The short lines mark the location of the Ridge crest, along which there is a positive anomaly. Despite the irregularities, the pattern shows a striking symmetry about the Ridge crest. From Heirtzler et al. [13], reproduced with permission

> To general astonishment, magnetic reversals provide the long-sought global stratigraphic markers that are revolutionising most of geology. At sea, as though by a miracle, magnetic anomalies give the age of the sea floor without even collecting a sample of rock.

Thus three very different kinds of evidence, seismic, sedimentary and magnetic, provided strong confirmation of the predictions of sea floor spreading and plate tectonics. By the early 1970s plate tectonics had become widely accepted, at first mainly by geophysicists but then by much of the broader geology community. A fundamental theory of Earth movements had been established for the first time.

Perhaps we can add the minor anecdote from Chap. 3, showing that rates of slip at subduction zones also fit the predictions, though with significant margins of error. Subsequently the *directions* of slip at many subduction zones have been established from seismic radiation patterns and found also to be consistent with predictions.

When Wilson read Dietz' seafloor spreading paper, he thought of using volcanic ocean islands as probes of the seafloor, reasoning that islands should be progressively older at greater distances from mid-ocean rises, and published his attempt in 1963. The idea was good, but the data were scattered and somewhat misleading, since some islands include fragments of continental crust, and other ages were not representative of the main phase of island formation. Even an accurate age for the main phase of building a volcanic island gives only a lower bound on the age of the sea floor upon which it is built. The sea floor will have the same age as the island if the island formed at a mid-ocean rise, but it will be older if the island formed away from the mid-ocean rise.

Despite the limited data available at the time, two clear ideas emerged from this part of Wilson's work. One was that some 'lateral ridges' could be explained if they represent the traces of extra volcanism at a mid-ocean rise. In fact, such volcanism might produce a complementary pair of ridges, one on each plate moving away from the mid-ocean rise. Wilson cited the Rio Grande Ridge and the Walvis Ridge in the South Atlantic as an example of such a pair, the active volcanism of Tristan da Cunha being the current site of generation. A closely related idea had been proposed by Carey in 1958, and Wilson acknowledged his debt to Carey. This is now the accepted interpretation.

The second idea was a mechanism to explain age progressions in island chains. Wilson recognised that there is active volcanism on some islands that are located well away from mid-ocean rises, so that some islands clearly had not formed at a mid-ocean rise, the Hawaiian islands being an outstanding example. With the idea of seafloor moving sideways, he realised that Dana's inferred age sequence for the Hawaiian islands could be produced if there was a (relatively) stationary source of volcanism deep in the mantle that had generated the islands successively as the seafloor passed over. He conjectured that this 'hotspot' source might be located near the slowly moving centre of a convection 'cell'. This work was also published in 1963. At the time Wilson conceived the idea the data on ages of the Hawaiian Islands were rather scattered, but also in 1963 more accurate dating was published by Ian McDougall of Australian National University, and it more strongly confirmed Wilson's interpretation.

W. Jason Morgan was a colleague of Hess' at Princeton University (Hess died in 1969). Morgan had already shown in a famous 1968 paper how Wilson's concepts of plates and fracture zones worked remarkably well when analysed in the spherical geometry of the Earth's surface. In 1971 Morgan turned to Wilson's hotspot idea, but proposed instead that there are *plumes* of hot material rising from the lower mantle. This hypothesis had a better physical basis than Wilson's. Morgan was familiar with analogous plumes in the atmosphere. The heat driving the mantle plume could emerge from the hotter core below the mantle. It was a common assumption at the time that the lower mantle, or 'mesosphere', has a very high viscosity and does not partake in convection, as depicted in Fig. 3.1, and so this would explain the relative fixity of the volcanism as the plate passed over the plume. Although this is not quite the current conception, the basic idea of a hot mantle plume has survived well. Morgan did not offer a sketch of his plume idea, but wrote of plumes as 'pipes' through the stiff mesosphere. He is also a bit vague about what drives plumes, as his focus was mainly on demonstrating the relative fixity of their surface expression.

The term hotspot is still used, but to refer to the *volcanic hotspot* at the Earth's surface, rather than to Wilson's 'mantle hotspot', an idea that was soon dropped in favour of a mantle plume.

Mantle plumes were not as quickly accepted as tectonic plates, though ultimately they have been a very fruitful hypothesis. Morgan actually proposed that plumes are horizontally fixed, reflecting his assumption at the time of a static mesosphere, and devoted much of his paper to demonstrating this. Hotspots do generally move more slowly than the plates, and hotspot fixity has been a useful approximation that has helped to refine details of plate motions, but it was taken too literally, as it was not long before the static mesosphere idea was dropped in favour of a lower mantle that is slower but not static. Morgan also argued that plumes are the major agent driving the plates, on the basis of a number of volcanic hotspots, like Iceland, that are on mid-ocean rises. This idea has not survived. The reality of plumes as a source of island volcanism became commonly accepted.

Later a rancorous debate developed, on the basis that some advocates claimed there was no good physical theory of plumes (which was no longer true) and that their presence in places like Iceland were not clear (Iceland is more complex because of the larger accumulations of eruptions there), and various other objections. That debate eventually faded, and we will look in more detail at how plumes work in a later chapter.

Wilson's work from 1961 to 1965 was exceptionally fruitful. In 1963 he published his papers on using islands to date the seafloor and on the age progression at Hawaii being due to passage of a plate over a (relatively) stationary volcanic source. He also published an extensive discussion of evidence for continental drift and seafloor spreading in *Scientific American* in that year. In 1965 he published his paper on transform faults and plate tectonics. This body of work brought far greater clarity to what we might call geological kinematics than had existed before. It also led ultimately to the identification of the two main dynamical sources of geological activity, a story we will pick up later.

A central puzzle that geologists had grappled with for two centuries or more was the cause of mountain building. Werner's theory was that mountain cores were left over from the early scouring by his great turbulent ocean. Hutton and Playfair presumed that mountains were thrust up in great convulsions, but did not have a clear mechanism for the convulsions. Lyell argued that mountains were uplifted slowly, perhaps as slowly as they were then worn away, but he had no clear mechanism to offer either. Others advocated thermal contraction of the Earth, tidal forces, centrifugal forces and even expansion of the Earth, but no hypothesis received general acceptance.

Wegener's idea of continental drift had clear potential to explain many mountain belts, yet it was too radical to meet with broad acceptance and there was no clear mechanism to guide inference anyway. With the advent of sea-floor spreading and its complementary subduction, and with Wilson's clarification of how those work, there was finally a theory capable of yielding testable predictions. A paper by John Dewey and John Bird in 1970 is an early example of how plate tectonics was seized upon to clarify the accumulated large body of geological evidence from mountain belts.

Figure 8.7 shows an expected geological evolution due to the collision of two continents, the one on the right being carried by a plate that subducts under the continent on the left. This is the kind of mountain building occurring in the Himalaya mountains at present. Because continental crust is less dense than oceanic crust, it does not subduct, but piles up at the former oceanic trench. Dewey and Bird argued

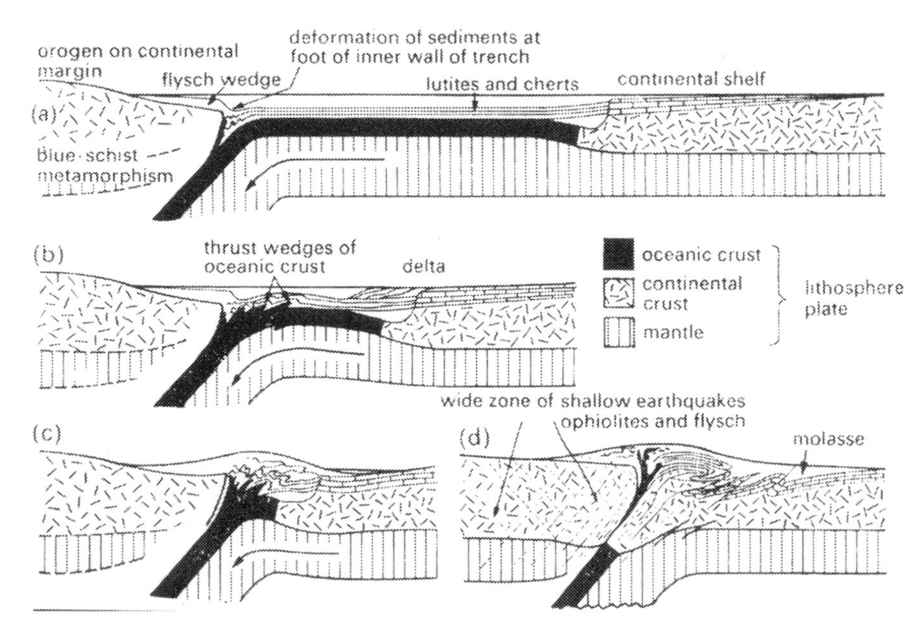

Fig. 8.7 Depiction of a sequence involving the collision of two continents. From Dewey and Bird [14], reproduced with permission

that encounters of rafted continents with various kinds of subduction zones could account for some of the main kinds of mountain belts.

There are endless details to explore and debate but, even without knowing the driving mechanism, plate tectonics provides a theory of mountains being raised, slowly and by inexorable forces, much as Charles Lyell had supposed.

There is additional novelty concealed within the seemingly simple description of plate boundaries laid out by Tuzo Wilson. Some of the novelty is evident in snapshot maps like Figs. 8.1 and 8.3. As I already remarked, the plates have a range of sizes and some odd shapes, some of them quite angular. However there is additional novelty in the way plates develop in time, so much so that it can be quite counter-intuitive.

These are the rules of plate motions, inferred from observation: the plates are rigid, and plate margins behave as follows.

- Spreading is symmetric, meaning equal amounts of new material attach to each of the plates that meet at a spreading centre.
- Subduction is asymmetric, meaning material is removed from only one of the two plates that meet at a trench.
- The relative motion of plates that meet at a transform fault is parallel to the transform fault.

Rather than presuming you know how things will turn out, you may need to step carefully through these deceptively simple rules. But if you do, the analysis can be quite powerful. This bears on how geologists use inferred plate histories to interpret their object of interest. It also bears on how plates relate to the mantle convection presumed to be their driving mechanism, which is the subject of the next chapter.

A striking example of the power of the rules of plate evolution came from the Indian Ocean, where the sequence of events has been rather complex. The outlines of the main phases of seafloor spreading were correctly inferred by Dan McKenzie and John Sclater in 1971 on the basis of a data set that was remarkably sparse for such a huge area. The Indian Ocean is much less travelled than the Atlantic, for example, so there were not a lot of magnetic surveys, and some of the data were from single 'ship tracks'.

Despite this sparsity McKenzie and Sclater were able to reconstruct the main phases of the evolution of the Indian Ocean, as first Africa detached, then India, then Australia from Antarctica. Given that there were four continents involved, and several distinct phases of seafloor spreading, this remains one of the more remarkable demonstrations of the power of the rules of plate motion. Their reconstruction showed India racing north at a blinding 18 cm per year, something of a land-speed record, before ploughing into the belly of Asia and pushing up the Himalayan Range in the manner of Fig. 8.7.

Even when plate velocities are constant and no new plate margins are forming, the sizes and shapes of plates can change. Plates can grow, and they can shrink, and even disappear. This is evident in the reconstruction of events in the Pacific inferred by Tanya Atwater in 1989 (Fig. 8.8) from the very busy-looking map of fracture zones and magnetic anomalies in Fig. 7.2.

Starting in panel B you can see the former Farallon plate between the Pacific plate and North America. Only a few small fragments of it remain. But then another extinct plate appears in panel C, the Kula plate in the north. This is required by the anomalies that turn east–west near the Aleutian Isalnds in Fig. 7.2. In panel F things look rather different, and yet another plate has appeared, the Izanagi plate. Thereby hangs a tale.

First, a bit of background. There were no magnetic field reversals between about 110 and 80 million years ago. This means the oceanic tape recorder was turned off, and sea floor formed during that time has no magnetic stripes to guide reconstruction. There are broad swaths of sea floor that are in this *magnetic quiet zone*. Older sea floor in the Western Pacific does have stripes, from field reversals that were happening prior to 110 million years ago, and they are called Mesozoic anomalies. However it was hard to figure out how to bridge the quiet zone and connect the plate motions before and after. If you just keep running the Pacific, Farallon and Kula plates backwards, as in panels C-E, Fig. 8.8, they do not match the Mesozoic stripes well at all. Something strange seemed to have happened in the quiet time, perhaps a large ridge jump or a very large transform offset, but these solutions seemed contrived and not very likely. Unlike in the Indian Ocean, the plate rules did not simply overcome a sparsity of data.

I had become fascinated by the magnetic stripes and reconstructions. In the course of teaching some graduate students about them, at Washington University in St. Louis in 1981, I noticed that the magnetic stripes near the so-called great magnetic bight in the north Pacific (Fig. 7.2) form a peculiar 'buttress' shape. It is shown in Fig. 8.9a,

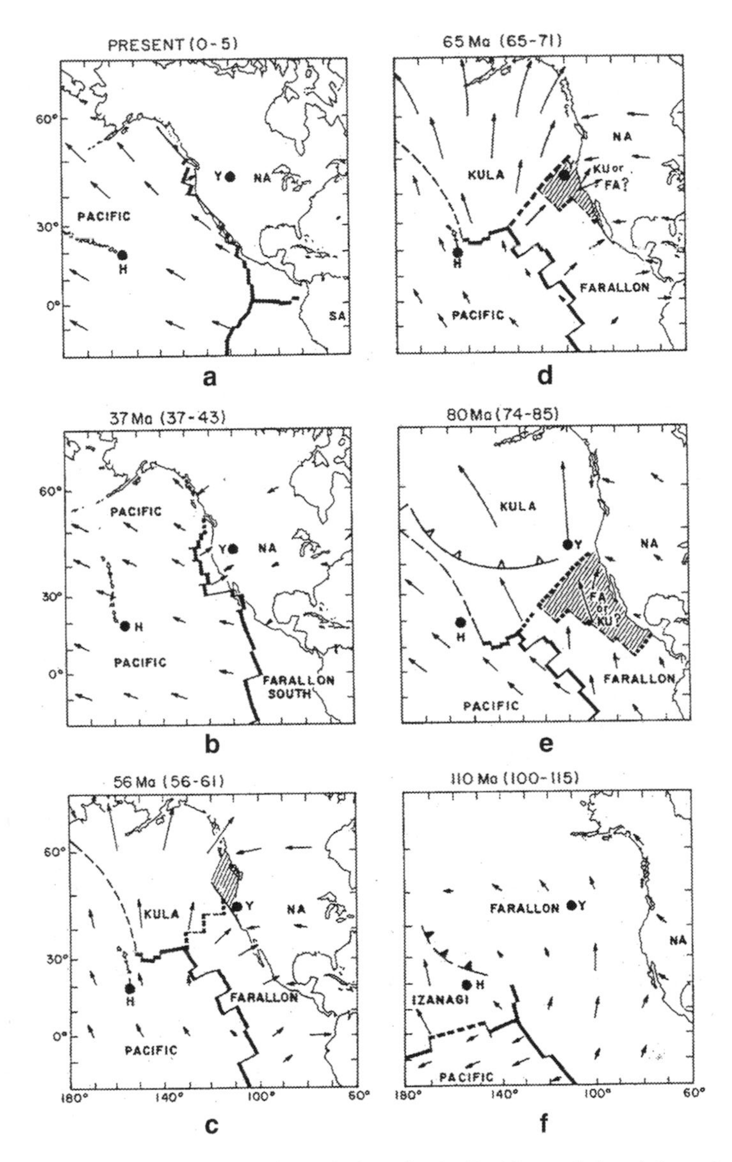

Fig. 8.8 Reconstruction of the evolution of plates in the Pacific, as deduced from the map of magnetic anomalies in Fig. 7.2. From Atwater [15], reproduced with permission

b. If you extrapolate this shape back in time it reaches an impossible configuration (Fig. 8.9c), in which a small piece of the Pacific plate would have had to pop up separately and then merge with the main plate at its northeast corner.

At my suggestion graduate student Mark Woods pursued the idea and developed the case that the Kula plate had actually formed by breaking off the Pacific plate about 85 million years ago, along what is now the Chinook fracture zone. Before that time,

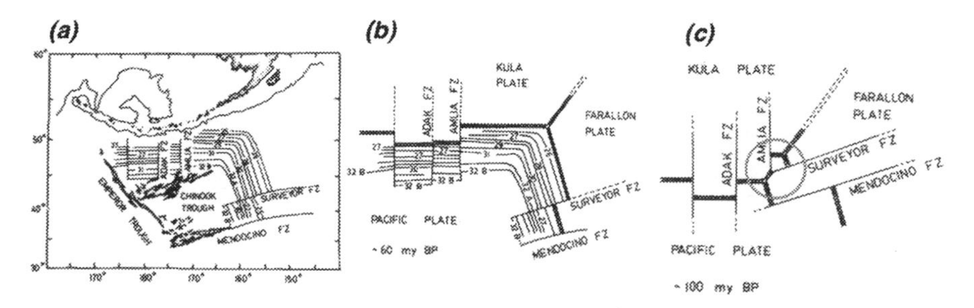

Fig. 8.9 **a** The 'great magnetic bight' in the northern Pacific, where the magnetic stripes turn from east–west to northwest-southeast, in relation to the Aleutian Island arc, the Chinook trough and other topographic features. **b** The inferred configuration of the Pacific, Farallon and Kula plates at the time of anomaly 25 about 60 million years ago. They form a *triple junction* (three plates, three spreading centres). **c** The plates at 100 million years ago if you just extrapolate back in time. This is an impossible configuration that would require another triple junction to pop up ahead of an existing one (circled area). After Woods and Davies [16], reproduced with permission

the Kula plate was just the northern part of the Pacific plate, and the impossible configuration of Fig. 8.9c never existed.

There was a big implication. The Mesozoic anomalies further west could not have involved the Kula plate, because the Kula plate did not yet exist. However there is another magnetic 'bight' within those anomalies, implying the presence of a third plate. If it wasn't the Kula plate it must be a different plate, previously unrecognised. It would have been moving northwestward and subducting under Japan. Mark proposed that we call it the Izanagi plate, after a being in Japanese mythology responsible for the creation of the Japanese islands. Perfect. The name also recognised a lot of exploration of the western Pacific by Japanese scientists. So that is how the Izanagi plate came to be in Fig. 8.8f.

We had inferred and named a very large piece of real estate, with ocean views. Unfortunately it is now somewhere down in the mantle under Asia.

These examples give an impression of how distinctively the plates move, growing and shrinking and changing, mostly following those simple rules. Even when they don't follow the rules, it is usually just by breaking a previously existing plate.

Subduction zones and transform faults correspond with standard types of faults identified long since by structural geologists, namely *reverse* or thrust faults and *strike-slip* faults. There is extensional faulting at mid-ocean rises, but the net effect is not like the standard *normal* fault type. The net effect is simply to pull the plates apart as new magma rises up in the gap and attaches equal amounts to each plate. This is illustrated in Fig. 8.10.

The fundamental point to emphasise here is that plates behave as a brittle solid, with the minor modification of how spreading centres work. Basically the plates are doing what the cold, brittle lithosphere wants to do. They do not seem to reflect much influence at all from a deformable, flowing, convecting mantle underneath. How can that be?

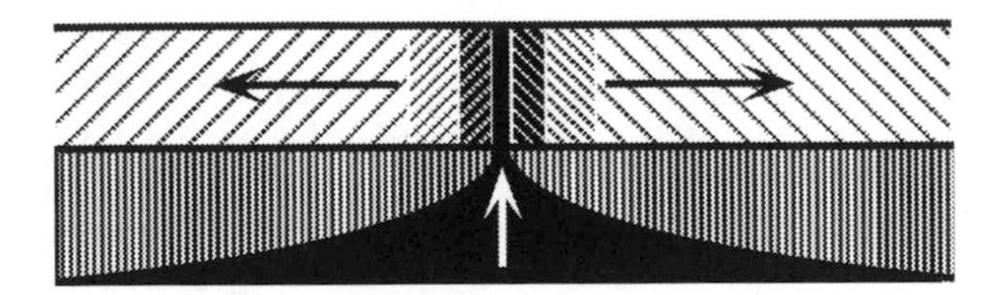

Fig. 8.10 Sketch of a cross section of a spreading centre at a mid-ocean rise. Magma rises at the centre, freezes and attaches to the previous crust. The process is observed to be symmetric in most cases. From Davies [4], reproduced with permission

Chapter 9
But What is the Driving Mechanism?

I once interacted with a colleague in the economics department where I worked in the US. He was a bit of a smart aleck, as quite a few economists are. He said, for example, that he could reduce everyone's heating bill to zero: 'Just take the roof off their houses.' He confidently presumed people would then turn off their heating systems. He gave me one of his recent papers to read. It comprised three parts. Part one was pretty much a statement of faith in free markets. Part two was some sophisticated, though old-fashioned, mathematical analysis of the kind economists are wont to do. Part three was a concluding statement of faith. I could understand enough of it to perceive a lack of connection between the middle part and the others. It was nice mathematics, but it did not seem to illuminate or justify the claims made in the other two parts. I wondered how it could have got past referees and an editor. Having later delved a lot more into mainstream ('neoclassical') economics I understand now that this is normal operating procedure. They have a theory of general market equilibrium, but it bears no useful resemblance to real economies. They carry on regardless.

I once attended a lecture by a prominent peer in my field. He drew quite a large crowd. He expounded on how mantle convection works and he showed his latest accomplishment, which was movies of three-dimensional calculations of convection in a spherical shell (like the mantle) – this would have been the early 1990s. He concluded with more comments on how mantle convection works. As the audience filed out I heard people remarking on how impressive the presentation had been.

I will not be so cruel as to say my colleague was no better than the economist. He was, at the least, using a computer to do calculations that cannot be done with analytical mathematics. Computers were just then reaching sufficient power to do three-dimensional numerical models. Accordingly his was a rather simplified model, with no plates and fairly low resolution. However I understood it was a step in a process, and that the models would get better, as they already had got better in two dimensions. Nevertheless I will say that he improvised rather freely around the computer movies. Evidently much of the audience was unaware of how free the improvisation was. His judgements about mantle convection were not outrageous,

G. F. Davies, *Stories from the Deep Earth*,
https://doi.org/10.1007/978-3-030-91359-5_9

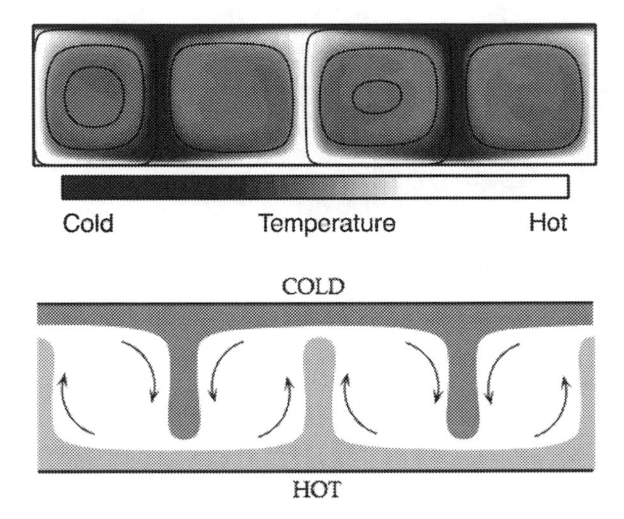

Fig. 9.1 **Top**: calculated temperature and flow lines in a gently convecting fluid layer, of the Rayleigh-Bénard type. **Bottom**: a sketch interpreting the top picture. Heat conducts in at the bottom, forming a hot thermal boundary layer that becomes buoyant and rises. Heat conducts out at the top, forming a cool, heavy thermal boundary layer that sinks. The rising and sinking columns drive 'cells' of circulation

it's just that they were judgements rather than conclusions from the computer models. Others, including me, had different assessments.

Although a three dimensional calculation was a step forward, the extra dimension used up so much computer power that the fluid had to be simplified back to the sort of thing we had been doing a decade or so previously in two dimensions. The trouble with that was that simple convection and plates tectonics do not much resemble each other.

Most thinking about convection among geologists and 'solid-Earth' geophysicists was conditioned by what is known as Rayleigh-Bénard convection. H. Bénard had, in 1900, done some careful experiments in which he gently heated a layer of liquid uniformly from below. The liquid began to overturn in a series of 'cells' that were about as wide as the liquid was deep. Lord Rayleigh in 1916 did a mathematical analysis that showed how Bénard's cells were a manifestation of instability in the layer.

Figure 9.1 is an illustration of this kind of convection. The convection is divided into a fairly steady pattern of cells, the cells are about as wide as the liquid layer is deep, and buoyant upwellings alternate with negatively buoyant downwellings. (I once looked for a word to complement 'buoyancy', for when a fluid is more dense and sinking. The best on offer seemed to be *ponderousness* or *ponderosity*. I decided to stick with 'negative buoyancy'.)

People commonly spoke of 'convection cells' and also of 'convection currents'. The latter term might be appropriate for the ocean, or for a molten mantle, but it never seemed to me to be a useful term for the creeping mantle. Nevertheless it is an indicator of the kind of thinking people were trying to bring to bear.

It is not clear how the kind of convection depicted in Fig. 9.1 might relate to the map of the plates and their velocities in Fig. 8.3. The plates are diverse in size: the Pacific plate is 14,000 km across, from the East Pacific Rise to the Japan Trench, whereas the Cocos plate near Central America is about 2,000 km across. The plates grow and shrink in unusual ways, as shown in Fig. 8.8.

The dissonance gets worse if you look again at Fig. 3.1. Isacks, Oliver and Sykes thought in terms of a 'return flow', from subduction zones to spreading centres, that balances the flow of plate material in the other direction. They supposed the return flow is confined to the *upper mantle*, above 650 km depth, because the *lower mantle* or mesosphere was believed to be so strong as to be immobile.

Actually in Fig. 3.1 they have shown the return flow even shallower than that, between about 100 and 300 km depth. They did that because seismic wave velocities in the upper mantle were known to go through a minimum there, forming what is known as the *low velocity layer*. If you recall, that is the depth range within which the mantle temperature approaches the melting temperature (Fig. 5.4). It is believed the mantle is close to melting there, and that this would cause both seismic velocities and the viscosity to be lower, and people often referred to the *low viscosity layer* (and still do). However there was not and is not much direct evidence for a pervasive low *viscosity* layer, except locally in some places. Even the low (seismic) velocity layer is confined mainly to sub-oceanic areas, and is muted or absent under continents. This made it difficult to understand how the continents might move as readily as the oceanic parts of the plates. Remember India, charging north at 18 cm/year, compared with an average plate speed of about 7 cm/year.

Even if the return flow (i.e. mantle convection) is confined just to the upper mantle, it implies convection cells that are very wide and flat. The average plate width is about 7000 km, more than 10 times the upper mantle depth of 650 km.

In 1976 Tuzo Wilson edited a collection of *Scientific American* articles entitled *Continents Adrift and Continents Aground*. In his conclusion he summarised a variety of ideas on the question 'What is the nature of the forces that move the pates about?' He notes the evidence that the rate of opening of the Atlantic had hardly changed in 80 million years and infers that the convection 'currents' must therefore be extraordinarily steady. Yet they must also have moved, because the mid-ocean rises wrap around Africa and were once much closer to the African continent and to each other. Indeed this was the evidence that had moved Bruce Heezen to propose expansion of the Earth. Wilson also notes the problem of the width of plates compared with the upper mantle or the low velocity layer.

Wilson mentions Jason Morgan's idea that mantle plumes might be the main drivers of plates. Morgan argued that volcanic hotspots like Iceland, and several others in the Atlantic Ocean, are surrounded by uplifts of the sea floor by as much as two kilometres and extending for some hundreds of kilometres around. These *hotspot swells* were presumed to be caused by ponding of hot mantle rising as mantle plumes. This amount of topography generates substantial gravity forces, so that the plates would tend to slide off the uplifts. In Morgan's conception the plumes are sites of rapid upflow which is balanced by a broad, slow sinking in the rest of the mantle. Whether the gravity forces were sufficient to drive the plates was not clear, and there are also a number of hotspots well away from the crests of mid-ocean rises, Hawaii being the outstanding example.

Anton Hales and Wolfgang Jacoby, reviving Hales' idea from 1935, had proposed that gravity sliding off the mid-ocean rises could drive the plates if the asthenosphere were sufficiently weak. This could yield the requisite wide, flat cells. Walter Elsasser of Princeton university had introduced the idea of the lithosphere as a *stress guide*, meaning that the tensional force from a sinking slab of lithosphere would propagate back into and along the attached surface plate, pulling it along after the sinking slab.

On the other hand Ernie Kanasewich, of the University of Alberta, suggested an underlying axial symmetry, not aligned with the spin axis, that might reflect a deeper pattern of convection. This was based on the two largest plates, Africa and the Pacific, being on opposite sides of the Earth with the other plates fitting between them. There *is* such a symmetry, though Wilson does not mention it: the gravitational *geoid* (the equipotential surface coinciding with the ocean surfaces and continued through the continents) has highs over those large plates and lows in a belt between them. Modern understanding is that this pattern is related to the plates but does not provide a primary driving mechanism, though it may reflect a subtle feedback from the mantle to the plates.

Wilson mentions proposals that there are two layers of convection, one in the upper mantle and a separate one in the lower mantle. The separation of these layers would require them to be compositionally different or, possibly, would require a strong blocking effect from phase transformations of mantle minerals at 650 km depth. Both possibilities were controversial.

A different approach to unravelling the driving mechanism, not mentioned by Wilson, was to postulate a set of forces and use the variety of plate sizes and situations to try to determine which are important. For example there was a ridge push force, a slab pull force, a basal resistance, transform fault resistance, and so on. Don Forsyth of Brown University and Seiya Uyeda of Tokyo concluded in 1975 that slab pull is the largest force but several others are important. This was an interesting approach but it suffered from ambiguities of interpretation. For example the slab pull was presumably a net pull due to the weight of the slab *minus* frictional resistance in the shallow fault zone where the subducting plate is in contact with the over-riding plate. However the magnitude of that resistance was not well known and it was the subject of a vigorous debate among seismologists. Resistance to the sinking slab deeper in the mantle might also be important, but was also poorly constrained. Similarly the small basal resistance they found could indicate a very low viscosity, shallow asthenosphere or that the mantle has higher viscosity but was rolling with the plate.

There were other conceptions of mantle convection as well. One was that there is a system of Rayleigh-Bénard cells under the plates and within the upper mantle. These cells would be driven by heat conducting from the lower mantle into the upper mantle. There would need to be multiple cells, about 600–1,000 km wide, under most plates. Principal exponents of such convection were Dan McKenzie at Cambridge University and Frank Richter at the University of Chicago. Richter showed, using laboratory experiments, that a plate moving over such convection could shift the plan-view pattern of cells from a square or hexagonal network to a set of rolls aligned with the direction of the plate above. Such 'cloud-street' rolls occur in the atmosphere on a windy day. These ideas generated quite a lot of work over the years, modelling this kind of convection and purporting to resolve gravity or topographic signatures of such cells or rolls, though the observational evidence always seemed to be marginal. Nor was it ever very clear whether these cells were driving the plates. That question seemed to have slipped into the background.

In many of these conceptions it seemed that convection was something that happened ‘down there’, below the plates for reasons not specified. This seems to be implicit, for example, in the sketch of Arthur Holmes (Fig. 7.3), though he did make some conjectures on how the convection was driven. The continents, or the plates, would then be passively rafted along on the top of these convection cells. On the other hand in Elsasser’s and others’ conceptions the plates played more of an active role. So this was another question: are the plates the active components or are they passive. It could of course be something in between.

Those versions that envisaged active plates with a shallow ‘return flow’ assumed or implied a low-viscosity layer under the plates that acted to ‘decouple’ the plate from the deeper mantle. Sometimes the idea of a decoupling or lubricating layer was stated explicitly. On the other hand the ‘active plate’ scenarios did not seem to offer an explanation for plates like South America that have no sinking plate attached yet still move.

This summary will give some idea of the confusion created by the contrast between the plates at the Earth’s surface and the convection presumed to be underneath. It was reminiscent of the confusion of the marine geologists as they revealed the strange features on the seafloor that don’t look much like the geology of the continents (Chap. 7).

It was around this time that I began to get involved again in plate tectonics and the driving mechanism. I had completed my PhD in 1973 and spent a couple of years as a postdoctoral fellow at Harvard University, mentored by Rick O’Connell. My PhD work was on elastic properties of minerals at very high pressures and temperatures, as exist in the deep mantle. Mainly I developed an integrated way of analysing data from various experiments, one being ultrasonic measurements of sound velocities and another being shock-wave experiments. My supervisor, Tom Ahrens, basically hung a small, innocent sample of a mineral at the business end of a naval cannon and fired a projectile at it. With very clever microsecond systems to record the event it was possible to extract the pressure, density and internal energy, from which one could also infer temperature. Incidentally, extracting the temperature required the use of the mysterious Debye model of thermal properties that I had encountered as an undergraduate (Chap. 2), so I finally found out what it was for. Tom’s recording technology was probably borrowed from the military too, used to study the effect of projectiles hitting targets.

The point of that work was to test various proposed compositions of the mantle by comparing their predicted properties with profiles of seismic wave velocity and density in the mantle, as summarised very briefly in Chap. 1. Thus I was well-placed to judge whether, for example, the lower mantle composition is different from the upper mantle composition. In fact part of my work was trying to see if any difference could be resolved. That was interesting because one of the chief proponents of a difference was the Director of the Caltech Seismo Lab, Don Anderson, who was one of my mentors. Don argued that the lower mantle was of a slightly different composition, enough to make it intrinsically denser so it would not mix back up with the upper mantle. That would require mantle convection to occur separately in the two mantle layers, as some were proposing.

There was a complication in this work, namely that the mantle minerals collapse into denser crystal structures because of the high pressure. This causes the density and elastic properties to increase. The result in the mantle is that the density and seismic velocities jump to higher values at around 420 km depth, and again at 650 km depth. This depth range is called the *transition zone*. It means you can't just extrapolate the properties of the upper mantle into the lower mantle, you have to take account of the extra increases due to these *phase transformations*. This makes it harder to constrain the lower mantle composition.

I concluded that the uncertainties did not require the lower mantle composition to be different from the upper mantle, though a small difference could not be ruled out. Don, on the other hand, thought you could resolve a difference. So I disagreed with one of my main mentors, and continued to disagree over some years. The debate was vigorous but civil, and it did not interfere with me getting my PhD.

My role as a postdoc at Harvard was to continue investigating mineral properties, including doing some high-precision ultrasonic laboratory measurements under high pressure. I can't claim to have accomplished a great deal of the experimental work, it was not my forté, but we did get some useful work done refining techniques and measuring a relevant mineral. As well, I found some systematic effects of phase transformations on elastic properties, using existing data, that had not been discerned before.

Rick O'Connell and I were both interested in mantle convection, and he undertook to run a 'seminar course' on the subject for graduate students, and he invited me in. Each week one of us would summarise an assigned paper for the class, and then we would discuss its lessons, and questions arising. Another participant was Brad Hager, who was one of Rick's graduate students. That class launched the three of us into the business of mantle convection. For me, it gave me the confidence that I was up-to-date on the important literature and would not therefore make a fool of myself by overlooking some important work. I might make a fool of myself in other ways, but not in that way at least.

Before going into our mantle convection work, I'll mention two other ways in which phase transformations are important. One way is that phase transformations may be triggers for the deep earthquakes that occur in descending slabs of lithosphere. The occurrence of earthquakes down to about 650 km depth was a puzzle. The cooler temperature in the slab would help but it was still not clear why the frequency of earthquakes would peak between about 400 and 650 km. The idea is that as the slab material sinks it decreases in volume because of the phase transformations, and that shrinking causes local stresses that trigger some kind of sudden slip, in spite of the very high pressures.

The other relevance of phase transformations is that they could, in principle, block vertical flow through the transition zone. For example, one of the phase transformations near 650 km depth is delayed in a slab because of the lower temperature, so

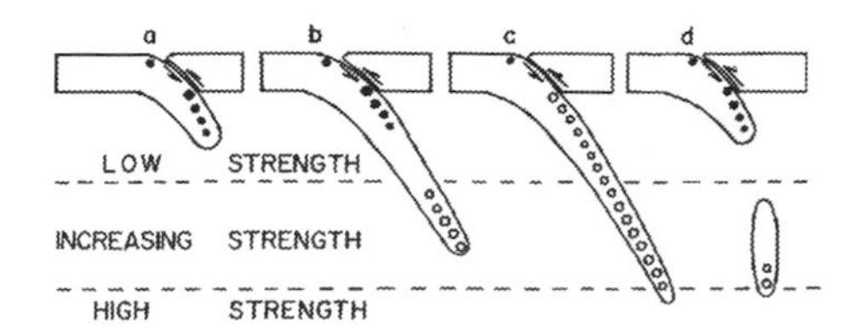

Fig. 9.2 Summary of stresses in descending slabs of lithosphere. Black dots indicate tension down the slab. White circles indicate compression down the slab, which indicates the slab is meeting resistance from below. From Isacks and Molnar [17], reproduced with permission

the material stays in its lower-density form as it descends into hotter material that is already denser. This exerts an upward buoyancy force that tends to resist the sinking of the slab. On the other hand the net effect around 420 km depth is probably to enhance the sinking. There was much debate over quite a few years. It seems the net effect is usually not strong enough to stop the sinking slab, though in some places it might. It might also have other effects that we will encounter later in this story.

For me, two insights in particular came out of our review of plate tectonics in Rick's class. One, stated concisely by Rick, is that the fact that deeply descending slabs are in compression does not mean they are stopped – I'll explain that shortly. The second was a paper that showed that a substantial increase of viscosity in the lower mantle would not be enough to confine the return flow to the upper mantle.

Rick and I both wrote papers over the next year or so, as I took up a faculty position at the University of Rochester in 'upstate' New York and Rick took leave to visit Cambridge, UK. Both of our papers were published in 1977. Fortunately they were nicely complementary. We each reviewed the evidence that flow through the transition zone is blocked by phase transformations or a change in composition, concluding that there was no compelling evidence for either possibility. We also addressed the possibility that the lower mantle viscosity is much higher than in the upper mantle. Rick discussed the micro-mechanisms for rock deformation in detail and the evidence from post-glacial rebound and other observations, concluding that a moderate viscosity in the lower mantle was plausible. I covered the latter evidence more briefly.

In a long paper in 1971, Bryan Isacks and Peter Molnar of Columbia University reported analyses of seismic wave radiation patterns from deep earthquakes. These patterns indicate the orientations of stresses that generated the earthquakes. They showed that some descending lithospheric slabs are in tension, as though they are being pulled downwards, whereas others are in compression, indicating resistance to their descent. There is a pattern in the occurrences of these stresses, summarised in Fig. 9.2. Where earthquakes extend down to only 300–400 km the slab is in tension. Where they extend to 650 km the slab is in compression. This was taken by many to reinforce the conclusion that slabs do not penetrate the lower mantle: they reach an obstacle at about 700 km depth and can sink no further.

Rick O'Connell, in our seminar course and in his paper, pointed out that the compression indicated in Fig. 9.2c does not mean the slab has stopped descending.

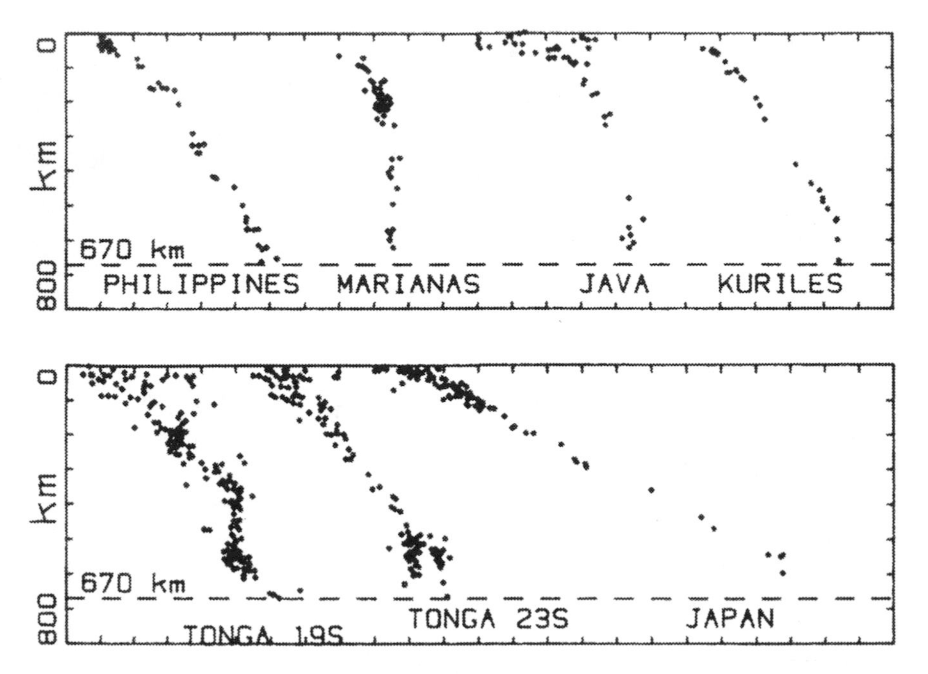

Fig. 9.3 Summaries of deep earthquake locations in the deepest Wadati-Benioff zones. From Davies and Richards [18], reproduced with permission

If you stand a brick on a table it will be in compression from its own weight, which is resisted by the table. If you stand the brick on quicksand it will also be in compression in the same way, but it can be slowly sinking. The compression indicates resistance, but not zero velocity.

I had also noticed that where earthquakes extended to the greatest depths, around 650 km, there is no indication of the slab being diverted horizontally, as it would need to be if it did not penetrate to greater depth. Figure 9.3 shows a summary of the deepest Wadati-Benioff seismic zones, which are taken to trace the descending slabs.

Only in the Tonga zone are there indications of contortion. The others all seem to slope smoothly down to their deepest parts, the Philippines being the clearest example. The suggestion is that the subducted lithosphere continues smoothly into the lower mantle. The earthquakes stop, but that does not mean the slab stops.

The cessation of earthquakes can be quite plausibly explained as due to the slab undergoing phase transformations in which the stress ends up being relieved. As the slab is also warming as it descends, it may never regain enough stress to cause more earthquakes. I took the shapes of the Wadati-Benioff zones to indicate that slabs penetrate into the lower mantle, but the idea that they indicate the slabs being stopped was firmly established and most people seemed to stick with that, despite the seeming inconsistency with the actual profiles.

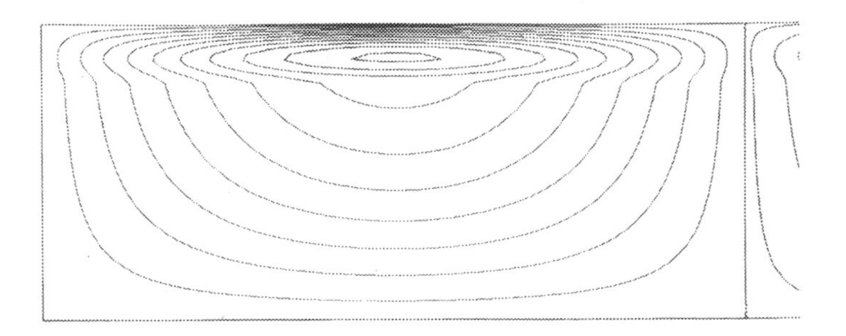

Fig. 9.4 Flow lines for incipient convection with layered viscosity. The lower 80% of the fluid has a viscosity 10,000 times greater than in the upper 20%. The fluid is assumed to be heated from below and cooled from above, just strongly enough to get convection going, as Bénard had done in his experiments and Rayleigh in his analysis. From Davies [19]

The second issue I wanted to address was raised by a paper by H. Takeuchi and S. Sakata in 1970 in which they calculated convection flow in a fluid whose viscosity was layered. The viscosity in the lower 90% of the layer was 1000 times greater than the viscosity in the top 10% of the layer. The 'return flow' occurred entirely in the lower layer, despite its much higher viscosity. They concluded that a broad cell much wider than the depth of the upper layer was possible, and would help to explain how the plates could be so much wider than the upper mantle is deep.

I wanted to explore other parameter choices, and also flow dragged by a moving upper boundary. They assumed the lower boundary to be no-slip (zero horizontal velocity), but the lower boundary of the mantle would be free-slip (resting on a liquid core). Their upper layer corresponded to the upper 300 km of the mantle, but a lower-viscosity layer occupying the upper 600 km would better correspond with the upper mantle. I also explored a range of viscosity contrasts. The flow that resulted from these changes is shown in Fig. 9.4 for the case where the lower layer is 10,000 times more viscous than the upper layer. Even in this case most of the 'return flow' is within the lower mantle.

Basically this shows that it is unlikely the lower mantle is static, and not participating in the flow associated with the plates. I also explored other calculations in which the flow is dragged by a moving upper boundary. This would be, in a sense, the opposite extreme of the convection calculations. Technically, the buoyancy driving the flow in Fig. 9.4 is distributed throughout the fluid. By imposing a moving boundary on a passive fluid I was in effect assuming the driving buoyancy is entirely concentrated within the plates, and the rest of the mantle is passive. The real mantle would be somewhere in between.

The result was similar, though not quite as dramatic. A viscosity contrast of at least a factor of 300 was required to exclude most of the flow from the lower layer. This version does not include any equivalent of a sinking lithospheric slab, which would tend to drive the flow deeper than the flow that is just dragged by the upper plate. Thus this case is as generous as possible to the possibility of confining the flow to the upper mantle.

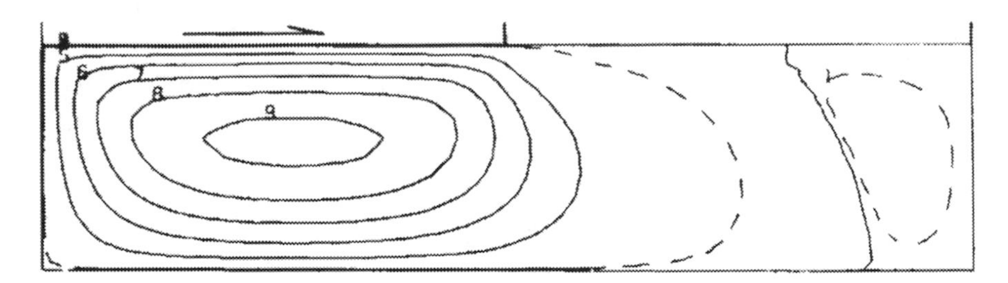

Fig. 9.5 Flow generated by a moving plate (left) converging towards a stationary plate (right). The flow descends obliquely under the 'subduction zone', and a weak counter-rotating flow occurs under the stationary plate. From Davies [20]

Another telling example had a low-viscosity layer between 60 and 120 km depth, to test the idea of a 'decoupling' or lubricating layer. The result was that the viscosity of this layer would have to be at least 10,000 times less before most of the return flow would be confined to it. This was, in effect, a test of the sketch in Fig. 3.1 depicting a return flow just under the plates, and it showed that situation to be quite unlikely.

I realised more could be done using this approach. Figure 9.5 is from a short follow-up paper looking at having two plates on the surface, one stationary and the other converging as it would at a subduction zone. The flow descends obliquely under the 'subduction zone', suggesting that the non-vertical inclination of real subducting slabs (Fig. 9.3) is due to the flow associated with the subducting plate extending for a distance under the over-riding plate, which is usually not moving very fast. The descending part of this flow also generates a weak counter-rotating cell under the stationary plate. This would tend to be amplified by a sinking slab, and could help to explain the motion of North and South America even though they have no sinking slab attached to them.

In the meantime Brad Hager was doing closely analogous calculations within a spherical shell, under the supervision of Rick O'Connell. They imposed the observed plate geometries and velocities on a passive, viscous mantle. In a 1978 paper they reported obliquely descending flows under subduction zones, like that in Fig. 9.5, that correlated well with observed Wadati-Benioff zones like those in Fig. 9.3, but only if the flow was allowed to penetrate into the lower mantle. If the lower mantle was made much more viscous the correlation with Wadati-Benioff zones was very poor.

In a 1979 paper, Hager and O'Connell showed that the flow due to the plates does not occur in 'cells' under each plate. Rather the flow is globally connected, with the strongest ascent under the southeast Pacific and the strongest descent under the Japan region, but the 'return' flow is quite indirect and does not just go from Japan back to the southeast Pacific. The absence of buoyancy forces in their models probably exaggerated this effect, but it was a further caution not to think in terms of simple Rayleigh-Bénard cells. This paper also confirmed in three dimensions that the lower mantle viscosity would have to be very high to exclude the flow, and that a shallow low-viscosity layer would have to be extreme for return flows to be confined there.

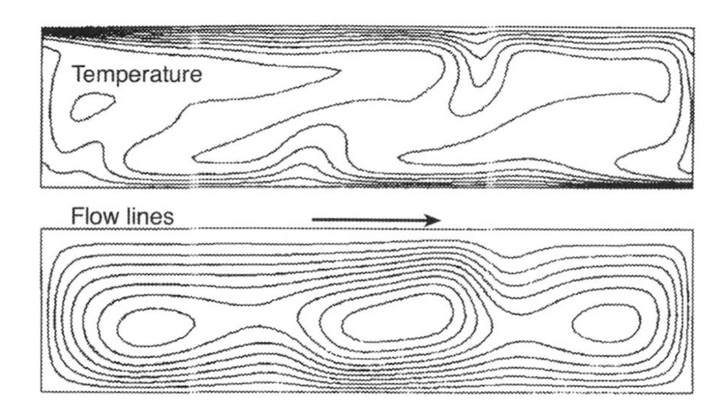

Fig. 9.6 An example in which the moving boundary on top is just able to impose a single long cell, but buoyant blobs are tending to break through it and make smaller cells. The top panel shows lines of constant temperature (isotherms), the bottom panel shows flow lines, with the direction of the motion of the top surface indicated by the arrow. From Lux et al. [21]

A next step would be to include thermal buoyancy forces along with a plate-like surface. After I moved to the University of Rochester I was approached by Rick Lux, an engineering graduate student who was interested in mantle convection. His engineering supervisor, Jack Thomas, was willing to share supervision, so I took Rick on. He developed a computer code to do thermal convection, which is rather more challenging than the 'kinematic' flows like that in Fig. 9.5. With it he was able to investigate the interaction between a moving plate at the surface and thermal convection underneath (in two dimensions). These calculations went beyond those like Fig. 9.4 in another way. In Fig. 9.4 the heating is just strong enough to start convection, but in our numerical models the heating could be rather stronger, as it is in the mantle.

The results were not too surprising, but Rick was able to quantify important details. If the plate was stationary or slow, you got Rayleigh-Bénard cells. If the plate was fast, you got a single, elongated cell, as in the left-hand side of Fig. 9.5. At the transition between Rayeigh-Bénard cells and a long single cell the flow was rather unsteady, as in Fig. 9.6. The picture now shows lines of constant temperature as well as flow lines. There is a cool layer at the top and a warm layer at the bottom. You can see a cool blob is starting to fall away from the top boundary and is deflecting the flow lines downwards. A warm blob is just starting to rise off the bottom as well. In this marginal case there is a competition between blobs tending to break through the flow and the horizontal flow tending to sweep them to the side before they break through.

The stronger the heating, the faster the blobs form, and the faster the plate has to move to maintain a single elongated cell. The results indicated that the mantle is just within the regime where the moving plate starts to dominate. Yet these models still lacked a realistic lithosphere, which would tend to make the plate more dominant by suppressing some of the unsteadiness under long plates.

It was looking like we were on track to resolve some of the puzzle about how large plates fit with convection. Yet there was clearly more to be done. A key feature of the plates is their internal stiffness, which is a result of them being cooler than the mantle interior. What was needed was to make the viscosity depend on temperature, so hotter fluid is less viscous and cooler fluid is more viscous.

The catch with that is it makes the computation much more difficult. With constant viscosity, the flow equation can be solved reliably and quickly on a computer. With variable viscosity you have to start with an approximate solution and progressively refine it. With each *iteration*, the solution should get better, in other words satisfy the flow equation more accurately. However if the viscosity variation is large the next iteration may not be much better than the previous one, so you need lots of iterations and it may be hard to determine if you have done enough. In the worst case the next iteration is worse than the previous one, and the solution 'blows up'. I don't recall exactly, but I must have recognised the difficulty, or just not had time to tackle it, being still a young faculty member trying to fit research around a busy teaching schedule. It was some years before I was able to take that next step.

After three years at Rochester the department was shrinking due to financial stringencies of the time. I was able to get a position at Washington University in St. Louis, Missouri, and moved there in 1978. I might also mention that my attempt to get more research funding from the National Science Foundation came to nought. My first NSF grant had supported Rick Lux's work, and although it was clearly going well my proposal was not found to be of sufficient merit. I don't think it is unreasonable to say the already intolerant culture in the mantle convection business, in a competitive funding environment, would have helped to sink my proposal. The 'whole-mantle' convection that some of us were arguing for was definitely a minority view at that time. The result was that I could not support Rick for the year or so longer that I would have preferred. He was able to get a postdoctoral position, but I don't think he was experienced enough to be properly independent. He was unable to get a faculty position, and went to work in industry. That's all right, but it's not what he wanted.

This is a story of how the science of mantle convection was done, not just a story of what was done. For that reason I have mentioned the funding aspect, and another factor may also have played a role. Most of the people I debated in mantle convection and, later, mantle geochemistry, had been launched into their work by a PhD supervisor already established in the field, or at least established at a prominent institution. Having the famous name as your co-author attracts attention and brings authority. My mentor in mantle convection was Rick O'Connell, who was only just starting in the subject himself, and we did not actually co-author anything on that topic. My PhD mentors were in other sub-disciplines. I was trying to climb into the subject under my own steam, and it does not seem to be the ideal way to proceed. Social and cultural factors play significant roles in how science progresses, despite the myth of 'objective science'.

There was other modelling of mantle convection being done during this time. Most models were tailored to the upper mantle, and many of them made little attempt to include anything resembling a lithospheric plate. Extensive modelling by McKenzie, Roberts and Weiss at Cambridge in 1974 is a prime example on both counts. A few studies did indicate that including stiffer plate-like behaviour could yield long convection cells, though they were specific to the upper mantle. Work by M. H. Houston and J. C. de Bremaecker in 1975 is an example, though at rather low resolution. Frank Richter's initial study in 1973 included a restriction on flow in the lithosphere equivalent to high viscosity, though there were other limitations of that early modelling. Mark Parmentier did an interesting model with temperature-dependent viscosity: it showed that making the cold fluid stiffer stabilised the lithosphere and allowed quite long cells, though with the flow confined to the upper mantle there was a significant horizontal pressure gradient that affected the calculated topography.

'Whole-mantle' convection was a minority interpretation, and there were certainly some who were dismissing it, airily or vociferously. Nevertheless my sense is that it was being taken reasonably seriously by enough people that it would have gained traction and, given the empirical success it was to achieve, it would have gradually been accepted.

However other influences entered, from another direction, and they quickly polarised opinion and raised the temperature of debates. Mantle geochemists proclaimed that the mantle comprised two layers of different isotopic composition. They must convect separately. Whole-mantle convection was not possible.

Chapter 10
Chemistry and Egos Muscle in

You knew you had made it in mantle geochemistry if Claude Allègre was shouting at you. The *prima donna* from Paris was a man of definite opinions loudly expressed. If it was a topic he considered worthy of his attention and your conclusion was at odds with his then you were clearly wrong. The most notorious occurrence was when Alan Zindler of Columbia University presented a new analysis of a set of isotopic data from mantle rocks. These were not new data, he merely used a three-dimensional plot to present them in a new way that revealed some interesting relationships. I was not there, but many reliable reports were that Claude occupied the whole question time (and then some) shouting the ways that Alan was wrong. But it wasn't Alan, it was the data that did not fit Claude's view of the world.

My turn came a bit later at a small, specialist conference in Turkey, in about 1987. The evening before, as people were arriving, Claude had asked me if I still believed that silly stuff about whole-mantle convection, or something along those lines. Once I had recovered from the surprise of having my existence acknowledged, I said yes. Disgusted that the world still did not understand the obvious truth, Claude persuaded the program organiser, a quiet, obliging person, to re-arrange the program and give Claude the opening speaker's slot—for two hours, several times longer than most other speakers. Claude went through all the usual arguments that I and many others had heard before. When eventually I got to speak the next day, for much less time, about how plate-related flow allowed isotopic heterogeneities to persist for billions of years, as observed, Claude was shouting from the back of the room " 'E knows nossing! 'E knows nossing!" It's amusing now, but back then I was still struggling to cope with such aggression.

I had been attending conferences for nearly twenty years by then, but they could still be challenging. A bit like a newborn, I took my early experiences as being just the way the world is, but I look back with some dismay. In my first few conferences some talks, mine and others', were received politely but others were aggressively criticised. The early experiences of aggression were a bit shocking, and it reflects badly on the culture of the community. That culture is aggravated by competitive funding, but it starts with fragile egos.

G. F. Davies, *Stories from the Deep Earth*,
https://doi.org/10.1007/978-3-030-91359-5_10

As a young graduate student you work on a project and it is exciting to get results that no-one has ever got before, even though they might not be earth-shattering. You try to get your material organised very concisely, because you will have 12 min to tell your story and 3 min of question time. Your turn comes, you talk into the darkened room, explaining your slides and results. You stumble a bit but get most of it out OK, and only a little after the buzzer. If you are fortunate there might be a couple of questions clarifying what you said. If you are less fortunate someone might stand up and rip into you and your results. Not being very expert yet, you stumble to respond. You feel betrayed. You were doing some real, original science and had some results that seemed to shed light. And someone just dumps on it.

As the socially awkward, nerdy foreigner from the colonies my reflex was to get defensive and be aggressive back. It didn't help of course. Nor did it help that some of the worst attacks, later, were from a senior colleague. I gradually acquired more self-confidence and a bit of poise and was more able to deflect attacks, but by then I probably had a reputation among some people. I suspect that followed me through my career, though I eventually mellowed a lot in my personal and professional life. But you see some people perhaps once a year in a conference setting and there is not a lot of time to know each other well. It is a strange, intermittent community.

In 1976 two groups reported measurements of neodymium isotopes in rocks from oceanic and continental sources. One group was Allègre's in Paris, the other was Gerald Wasserburg's at Caltech. There was a fierce rivalry between the groups. People had been measuring isotopes in rocks derived from the mantle for some time, notably of strontium and lead. A distinctive feature of neodymium is that meteorites provide a benchmark isotopic composition that is reasonably inferred to be the average composition of the Earth. Such a benchmark is not evident for lead or strontium. The Paris and Caltech groups found some oceanic basalts are different from the meteorites and some continental basalts are similar to the meteorites. Don DePaolo, Wasserburg's student, and Wasserburg published a second short paper in 1976, compiling data from both groups, and produced the graph in Fig. 10.1.

We don't need to go into the obscurities of this graph (explained briefly in the caption). A key point is that the circles are mostly not on the horizontal line and the squares are mostly near the horizontal line. The horizontal line corresponds to the benchmark meteorites (called *chondrites*). The data suggest that the continental basalts have come from sources in the mantle whose neodymium is close to the meteorite value. This *could* be because those sources are *primitive*, meaning they have not been melted or otherwise altered in any way since the Earth formed. Most of the oceanic basalts are clearly not on the meteorite line. The sloping line is a suggested correlation that would tell us about the relative abundances of these elements.

The authors suggested the data could be explained by two mantle source types, one primitive ($\varepsilon = 0$) and one up around+12 on the plot. The data spread along the sloping line would represent mixing of these two '*end-member*' source types.

I show this plot to emphasise the slender evidence upon which an interpretation was built. The data are sparse, a total of 7 oceanic and 8 continental. There are continental data (squares) at -4 and two at + 6 (partly obscured), well off the meteorite line. The point at -4 cannot be explained by mixing sources at + 12 and 0. There are

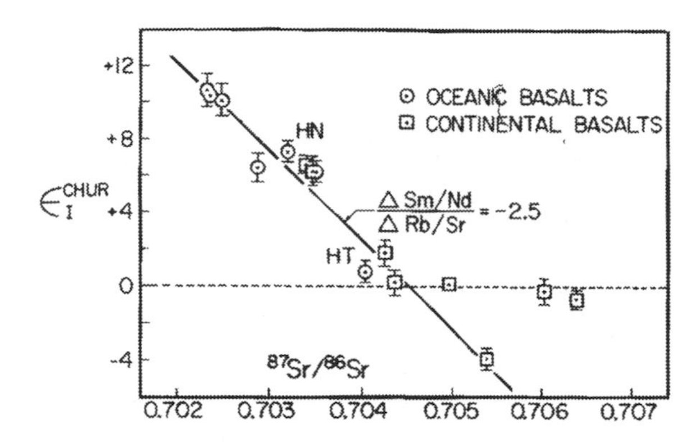

Fig. 10.1 Neodymium *versus* strontium isotopes from oceanic and continental sources. The strange symbol on the vertical axis represents deviations of the neodymium isotopes from the meteoritic reference value. Sr means strontium, and the horizontal axis is the ratio of two of its isotopes, of atomic weights 87 and 86. The vertical axis is another ratio, of neodymium 143 and 144; (ε represents deviations of the ratio in parts per 10,000, *I* means initial, CHUR means chondritic uniform reservoir, and chondritic refers to the kind of meteorite that provides the best benchmark for the Earth. Got it?). Sm (samarium) and Rb (rubidium) are 'parent' elements with isotopes that decay into Nd and Sr isotopes, respectively. From DePaolo and Wasserburg [22], reproduced with permission

also two continental data well off the correlation line to the right. These also cannot be explained by mixing the two proposed sources. The authors allow these might have been contaminated in their passage up through the continental crust, which has much higher strontium ratios.

One can do a 'mass balance' of the relevant elements. For example, they calculated that the continental crust contains about one-third of the Earth's complement of neodymium. As it happens the upper mantle comprises about one-third of the mass of the mantle, but there is not much neodymium in the upper mantle: *voilà*, perhaps the neodymium in the continental crust has come from the upper mantle. *If* the lower mantle were primitive, that would account for all the neodymium in the Earth. On the other hand if a presumed deeper reservoir were not primitive, then more, or less, of the mantle would have been depleted of neodymium. Evidently the correspondences (coincidences) were too irresistible, despite the questions, uncertainties and sparse and misfitting data.

More data were forthcoming in due course, but in 1979 Wasserburg and DePaolo (in that order) published a more explicit interpretation. They presumed two or three kinds of source in the mantle. After considering possible geometries for reservoirs containing the two main sources they produced the picture of the mantle shown in Fig. 10.2. In coming to this picture, they would have been aware of the opinion of many geophysicists that the lower mantle does not partake in the motions related to plate tectonics, and the picture of continents and lithosphere is very much the plate-tectonic picture. On the other hand papers showing the lower mantle probably has a moderate viscosity, insufficient to stop it, had been out for a decade or so.

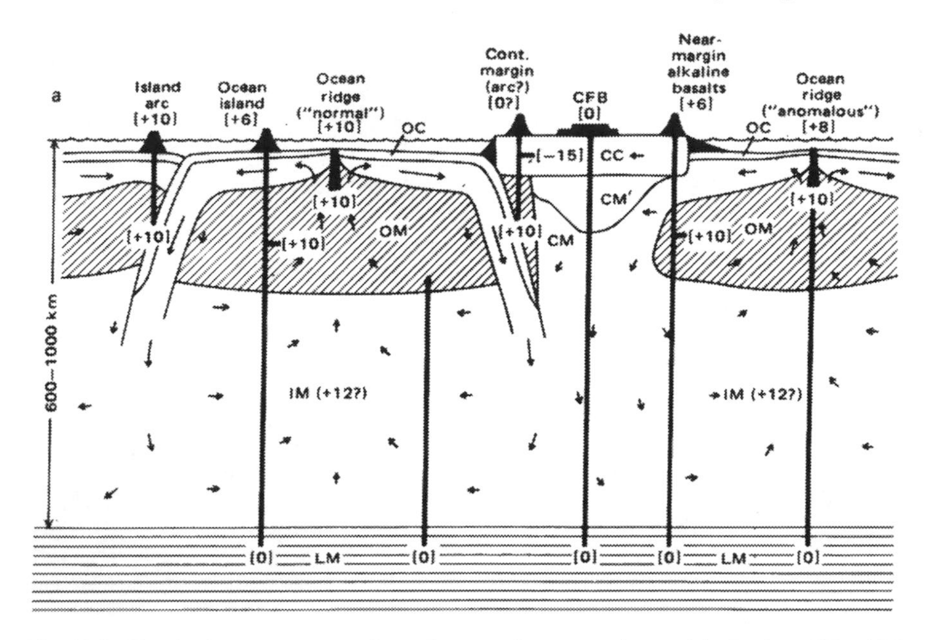

Fig. 10.2 Sketch of an arrangement of mantle reservoirs proposed to explain the data of Fig. 10.1. LM: lower mantle; IM: intermediate mantle; OM: oceanic mantle; OC: oceanic crust; CC: continental crust. Numbers are presumed values of ε for neodymium. From Wasserburg and DePaolo [23], reproduced with permission

The lower mantle (LM) is presumed to be primitive. The 'intermediate mantle' (IM) is presumed to be strongly processed and strongly depleted of neodymium and its parent samarium. The region tapped by mid-ocean rises (OM) is presumed to be slightly less 'depleted', through contamination by plumes passing through from the lower mantle; it is close to melting and more prone to such contamination than the intermediate mantle. Some primitive material makes it to the surface in continental flood basalts (CFB), whereas other primitive material picks up some depleted material on the way up.

Most layered versions of the mantle did not make a distinction between IM and OM. The Paris group came to a fairly similar picture, with a 'depleted' upper mantle and a primitive lower mantle.

Another interpretation of the data in Fig. 10.1 (and later versions with much more data) is that the continental basalt data are contaminated by continental crust, which is rather heterogeneous but generally has higher strontium ratios and more negative ε-neodymium. Regardless of possible contamination, one could also conclude that the mantle is rather heterogeneous, and that the spread of data in Fig. 10.1 simply reflects that diversity.

Before proceeding I want to acknowledge that Don DePaolo was a student at this stage, and obviously strongly guided by his supervisor Wasserburg. I later found Don to be reasonable and open in debate. He even acknowledged in conversation at one stage that he had learnt from me that chemical data are affected by physical processes as well as by chemical processes, and you need to take account of both. (Hallelujah! Trying to educate geochemists was not a very rewarding task, though it was more rewarding than trying to educate economists, as I have attempted more recently).

I knew Gerry Wasserburg from my first days at Caltech as a graduate student. He taught part of a crash course in geology for people like me who had little or no geological background. He could be a bit volatile, and throw chalk at someone not paying attention or not thinking clearly, but he was generally a careful, thoughtful teacher anxious to convey important ideas. On the other hand he did have a sharp tongue that was provoked sometimes. Gerry was more urbane than the bombastic Allègre, but we should have no doubt there was a strong ego present. It emerged that he could be quite inflexible in his interpretations. He stuck with the layered mantle long past the time when there was strong evidence against it, and had students do projects and theses on that basis, which I thought was not very responsible.

I also acknowledge that both Gerry's and Claude's large accomplishments were to establish innovative laboratories that generated many of the important data that we then argued about. I have already remarked in previous chapters that ultimate interpretations do not always come from observers and measurers.

The conclusion from early neodymium isotopic evidence that the mantle has two distinct layers was a convenient fit with the static or separate lower mantle still envisaged by some geophysicists, and it strongly reinforced opinion to that effect, despite the rather limited evidence base. However this picture did more than reinforce the layered mantle picture, it introduced and entrenched three key geochemical conceptions.

- *Reservoirs.* The array of data may suggest two *source types*, 'depleted' and 'primitive', but these were conflated with physical reservoirs in the mantle.
- *Uniform* reservoirs. Uniformity is explicit in *CHUR*, the chondritic uniform reservoir. But uniformity is implicit too, because if the upper mantle and the lower mantle did not each have relatively uniform composition what would be the point of distinguishing between them?
- *End members.* The spread of data is interpreted as resulting from *mixing* between material from the different reservoirs, so the reservoir compositions are necessarily at or beyond the extremes of the data. They are hypothetical end members.

I remarked in Chap. 3 on the coloniser effect in scientific ideas. In that case it established in many minds the static or separate mesosphere as a limit to the mantle convection associated with the moving plates. An even stronger coloniser effect occurred here. Not only was the two-layer mantle proclaimed as being definitively established, but thinking about the range of mantle isotopic and trace-element compositions of all kinds was almost universally framed in terms of extreme end-member compositions residing in uniform reservoirs.

The end-member framing continues strongly to this day, particularly regarding the composition of the upper mantle tapped by mid-ocean rises, the so-called *MORB source*, that is, the source of the *mid-ocean ridge basalts* that comprise most of the oceanic crust. There is, allegedly, a reservoir called *DMM*, the *depleted MORB mantle*, generally taken to be the uppermost mantle, although its vertical extent is often left unspecified.

There are certainly MORB samples that reflect a source strongly depleted in fusible (more easily melted) components, with the depletion reflected in so-called major elements, mineral components, trace elements and isotopes. But are those 'depleted' samples from a large-volume, contiguous reservoir or are they just the less-common tail of a dispersed distribution of source types that are generally not as depleted? The DMM concept channelled thinking about mantle structure, and also distorted estimates of mean composition of the mantle, which relates to fundamental things like the amount of radioactive heating in the mantle and the relation of the Earth to meteorites.

The main 'depletion' referred to here reflects melting of mantle that ascends under mid-ocean rises and melts as it approaches the surface. The basaltic magma so generated erupts at the crest of the rise to form new oceanic crust, and it leaves a residue depleted of those components that went into the melt (it's a complicated process, don't ask). Melting in other settings causes other changes with different signatures. Melting at subduction zones, forming volcanic island arcs or continental margins, is believed to further extract components that end up in continental crust through additional complicated processes.

The oceanic lithosphere is stratified because of melting at mid-ocean rises, with basaltic crust on top and strongly melted and depleted residual mantle rocks below. There is usually also some continental sediment accumulated on top by the time it reaches a subduction zone, some of which seems to make it into the mantle.

Most geochemists would probably agree that subduction of this thick sandwich introduces heterogeneities into the mantle. They have generally been assumed to be homogenised by the action of mantle convection. That is one question that was to be addressed in later studies.

It is also well known that there are variations in the composition of mid-ocean ridge basalts, usually reflecting a source that is less depleted or even enriched in the components extracted into the crust. The different kinds are given names like E-MORB, for 'enriched mid-ocean ridge basalt'. The parts that are not obviously different are called *normal MORB*.

People would remark on 'the remarkable uniformity of MORBs', reflecting both the steadiness in time and space of melting at mid-ocean rises and also the presumed uniformity of the MORB source. I began to wonder how, exactly, you defined 'normal'. If there were a clear bi-modal distribution of compositions, 'normal' in one mode and 'enriched' in the other, then it would be clear. The question did not seem to be clearly addressed, but I gathered there was no such obvious separation. Much later, in 2006, Garrett Ito and John Mahoney of Hawaii plotted histograms of strontium, helium and lead isotopes, and the strontium data are shown in Fig. 10.3. Even here the MORB data have had the more extreme date excluded (any rise crests above 2000 m and below 4000 m), but it is clear there is only one peak. It is skewed but there is no suggestion of bimodal distributions for either MORBs or OIBs.

Some people resorted to proxies. 'Enriched' MORBs tend to occur where the crest of a mid-ocean rise is higher than normal. Sometimes there is a plume nearby that

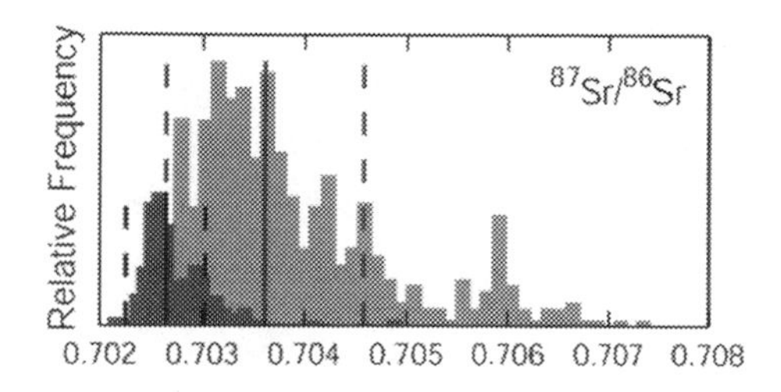

Fig. 10.3 Distribution of strontium isotope ratios for mid-ocean rise basalts (MORBs, dark grey) and ocean island basalts (OIBs, light grey). From Ito and Mahoney [24], reproduced with permission

might explain that, but not always. One time I was giving a talk at Cornell, making the point that it wasn't clear how you separate 'normal' from the rest. Don Turcotte, the resident guru of mantle convection, interjected to say 'normal' is when the rise crest is below a certain depth in the ocean. If you look at the distribution of depths of rise crests, you don't see any clear division there either.

It was beginning to look as though the correct statement is 'Normal MORB is remarkably normal'. In other words the logic was circular: you exclude samples that deviate beyond a certain narrow limit, then exclaim over how uniform the remaining samples are. I think it is probably true that this thinking was channeled by an expectation that there is a uniform reservoir in the upper mantle, notwithstanding the well-known diversity of what actually comes up at mid-ocean rises, for which special explanations were offered. The 'uniform reservoir' meme was doing its work.

And then there was the primitive reservoir, allegedly residing in the lower mantle. It had been known for some time, due to pioneering work by Paul Gast and others, that the lead isotopes of oceanic basalts are quite diverse. That diversity required the source of oceanic basalts to have been processed (read 'melted') in the more distant past. If the source has been processed it would seem it is not primitive.

I once asked, at a conference, how come the lead isotopes in the rocks from 'primitive' sources were not themselves primitive. The answer was that the lead isotopes were 'decoupled'. I had heard the term being tossed around but no-one seemed to explain what the decoupling process was. Perhaps they implied that diverse lead signatures were added at some later stage without disturbing the rest of the rock's primitive character. As there was clearly some variation also in other isotopic systems, like neodymium and strontium, perhaps the rocks were not really from a uniform primitive ('chondritic') reservoir. But people did not seem to want to admit that into their thinking.

It is sobering to look back at those times and try to reconstruct and summarise the arguments as they stood. The geochemical papers are quite dense. The jargon was highly specialised. There are many elements and many isotopic systems. Figure 10.1 is a taste of the specialist nature of the topic. Most notably in retrospect, the literature is large. There was a steady stream of papers reporting new chemical data, many details of hypotheses were debated, and apparent conflicts of data, interpretations and opinions abounded. I can only mention a small sample of key papers and points here.

Most geophysicists could only take most of the geochemistry at face value. By around 1980 I was at Washington University with a number of geochemist colleagues.

They were not directly involved in the mantle debates and I could discuss the topic with them and clarify my understanding. Those of us arguing for whole-mantle convection were of course bothered by the incompatibility of the geochemical conclusions with our growing understanding of mantle convection, which really did not work well if the plate-related flow is confined to the upper mantle.

In 1981 I published my first foray into mantle geochemistry, in *Nature* magazine. I was looking at the uncertainties underlying the layered mantle interpretation, and whether an alternative interpretation might possible. A key question was how much of the mantle would have been depleted to form the continental crust, and the uncertainties clearly allowed for up to 50% of the mantle to be depleted, or even 65%. In that case the identification of the depleted source with the upper mantle would not work.

There were some recent data indicating the continental 'keel' (the strong, attached mantle underlying continental crust) contained enriched material, though the amount of it was not well constrained. That would require more of the rest of the mantle to have been depleted. This possibility has not been pursued, probably because it is hard to constrain the amounts, but it strikes me as still quite plausible.

Another argument was that if one allows a component or reservoir within the mantle that is enriched rather than primitive, as some evidence was indicating, then even more of the remaining mantle would need to be depleted. I think this remains a robust argument.

One of the reasons it is tricky to pin down the exact arguments early on is that the 'continental' data in Fig. 10.1 ceased early on to play any role in the discussion. They had been the basis of the claim of a primitive reservoir, but they were dropped, presumably because it became clear they might well have been contaminated as they rose through the continental crust. Their neodymium signature would then have been only coincidentally primitive. I am not aware of any explicit reason being given for their disappearance. Later data focussed mainly on *oceanic island basalt* (OIB) as a complement of mid-ocean ridge basalt (MORB). We will explore OIB shortly, and find there is diversity among them, with not much reason to invoke anything primitive.

Another point advanced in my paper is that the correlation of neodymium and strontium isotopes, as in Fig. 10.1, might directly represent heterogeneity in the mantle rather than mixing between two end-member reservoirs. That remains a fundamental point that we will eventually return to.

The paper proposes a heterogeneous mantle convecting throughout its depth. The sketch of the concept looks a bit rudimentary now, but it was a step in a direction that became clearer over the next few years. The paper points out that heterogeneities are introduced at subduction zones and identifies their survival time as a key question. It distinguishes the *survival time* of a heterogeneity, which would depend on whatever shearing and distortion it experienced, from the *residence time* of mantle material, which is the time between passages near the surface where it might melt. The latter was estimated to be around 4 billion years at present rates. These are concepts and themes that guided much later work.

A general point is that examination of the geochemical data and their uncertainties revealed that alternative interpretations might be possible. This also was to be a consistent theme in later debates.

Uranium has a number of *isotopes*, which means there are versions of uranium with different atomic weights. To be technical, the different isotopes have different numbers of neutrons in their nucleus, but the same number of protons. So for example there are ^{235}U and ^{238}U, where the number is the atomic weight, which is the sum of the number of protons and neutrons. Some of the uranium isotopes undergo radioactive decay, which means bits of the nucleus fly off randomly, but at a very steady rate on average, leaving an atom of a different element. The decay releases a little chunk of energy, which is why you have heard so much about uranium, because if you get them all to go off at once you get quite a big bang. Actually uranium goes through a series of decays, ending up as lead. ^{235}U decays into ^{207}Pb and ^{238}U decays into ^{206}Pb. (Pb is the symbol for lead, which is why you get a *plumber* to fix your pipes, which used to be made out of lead until people got sick of lead poisoning messing up the Emperor's brain, and sanity.)

I could have explained this before, but you could get the essence of Fig. 10.1 without making it even more complicated. So let's now mention that ^{87}Rb (rubidium-87) decays into ^{87}Sr (strontium-87) and ^{147}Sm (samarium-147) decays into ^{143}Nd (neodymium-143). The radioactive isotope (e.g. ^{87}Rb) is called a *parent*, and the product (e.g. ^{87}Sr) is called a *daughter*. The daughter is also described as being *radiogenic* – so the radiogenic daughter is born of the *radioactive* parent. It is convenient for geochemists to refer parent and daughter to a *stable isotope* (e.g. ^{86}Sr) that is *non-radiogenic*.

I once went to hear a talk by Claude Allègre at a big conference. He thought only specialist geochemists would be there. He bragged that every time the geophysicists understood one radioactive decay system the geochemists would measure new one. I'll spare you the French accent this time. Anyway don't blame me for all these complications, blame him. Of course it means I'm about to hit you with more isotope diagrams.

It is also time to introduce the *noble gases*, briefly. We'll do more detail on them later. These are the most unreactive elements, they don't mix with the hoi polloi, hence the term *noble*. Sometimes they're called the *rare gases*, but argon makes up one percent of the atmosphere so that's not really appropriate. They are helium, neon, argon, krypton and xenon, respectively He, Ne, Ar, Kr, Xe.

Briefly for now, there should be more argon in the Earth than we can account for in the atmosphere, crust and upper mantle. It's called the missing argon problem. Also some ocean island basalts have helium that looks more primitive than average. By that is meant that the ratio of its two isotopes, $^{3}He/^{4}He$, is higher than usual. ^{4}He is a by-product of the decay of uranium and is thus radiogenic, whereas ^{3}He is a non-radiogenic isotope. ^{3}He is therefore called primordial or primitive, and a high ratio $^{3}He/^{4}He$ is taken to indicate 'more primitive'.

Figure 10.4 **is a summary** from 1997 of isotopic data from various kinds of oceanic basalts. MORB from the Pacific, Atlantic and Indian Oceans is there. There are also data from what geochemists call Ocean Island Basalts (OIBs), from volcanic

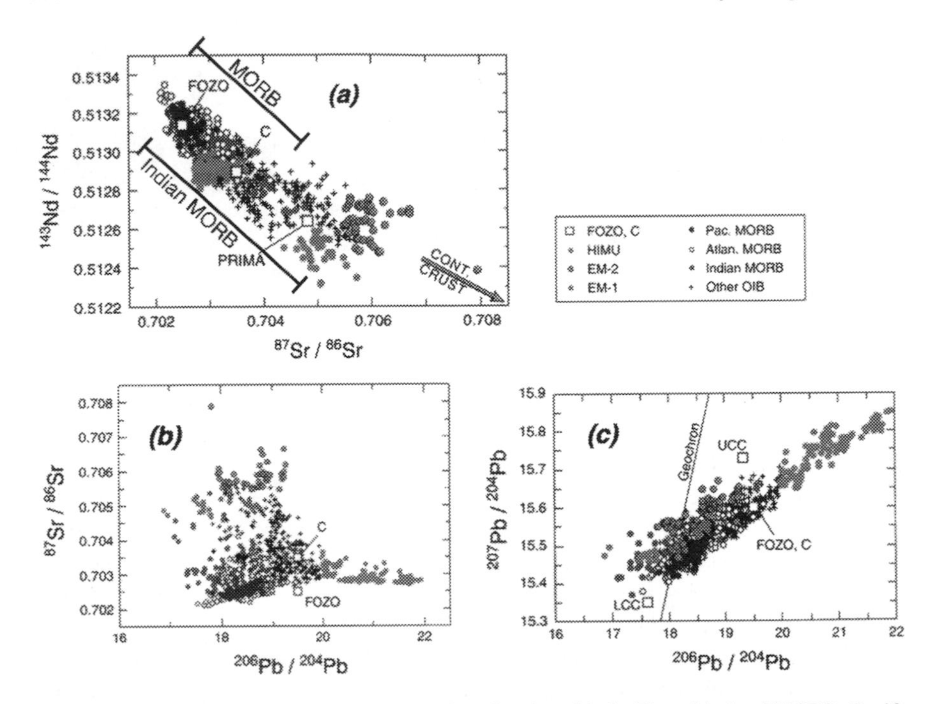

Fig. 10.4 More isotope plots, for oceanic rocks of various kinds. Three kinds of MORB: Pacific, Atlantic and Indian Ocean. Four kinds of oceanic island basalts (OIB: HIMU, EM-1, EM-2, Other). Three proposed mean compositions (FOZO, C, PRIMA). Continental crust, upper (UCC) and lower (LCC). After Hofmann [25], reproduced with permission

islands such as Hawaii and Iceland and quite a few other places. These are the islands that Tuzo Wilson identified as the product of hotspots, and Jason Morgan proposed instead are the product of hot mantle plumes rising under the plates from deep in the mantle.

For the moment look at panel (a). The original was coloured and clearer, so I have marked the extents of the MORBs. The Atlantic and Pacific MORBs are near the top left of the array, although the Indian Ocean data spread down past PRIMA. PRIMA (primitive mantle) marks the supposed primitive composition, identified from neodymium isotopes in primitive meteorites, and most of the MORB data are well up the array from that point, in what we can call the *depleted* quadrant. Many of the OIB data are also in that quadrant, though not so far up, and some extend into the lower right or *enriched* quadrant. The OIB also spread over a fair range of the strontium isotopes.

It should be evident that it is hard to explain the data in panel (a) as due to mixing between a PRIMA component and a depleted end-member component in the top left corner. A more accurate description of the data is that the MORBs are spread in the depleted quadrant but extending down near PRIMA, whereas the OIBs spread over a greater range from less depleted to slightly enriched, with less correlation.

Panel (b) plots strontium against a lead isotope from the same rocks and you see essentially no correlation and no hint at all of anything you could call a mixing line. Again the OIBs spread over a greater range than the MORBs, but neither shows a simple pattern. These are the kind of lead data that had been dismissed as being 'decoupled' from the neodymium and strontium data.

These plots are from a later and more complete compilation, but such data were accumulating through the 1980s and to some people they showed little evidence of a primitive mantle component. Nevertheless the primitive lower mantle reservoir was vigorously defended, in part because of the noble gas data. If the lower mantle were primitive, then it would contain extra argon and its helium would be more primitive, so both the noble gas questions would be resolved.

Even so, maintaining the two-layer mantle in the face of data like those in Fig. 10.3 was a challenge. It was allowed that some of the data showed signs of some input from, for example, continental crust, and it was allowed that this could be from sediments that go down with subducted lithosphere. These would then mix at the bottom of the upper mantle, it was claimed, with primitive material leaking from the lower mantle. Geochemist Claude Allègre and geophysicist Don Turcotte teamed up in 1985 to argue for such mixing in what they called the *mesospheric boundary layer*.

This was the context in which three people looked afresh at the data and proposed a quite different interpretation. Al Hofmann and Bill White, then at the Carnegie Institution of Washington argued, in 1980 and 1982, that the OIBs contain more *incompatible elements* than could be explained by just melting normal mantle, in the way MORBs are produced. Their isotopic compositions indicated the same thing. Incompatible elements are those that don't like being confined in the dense minerals the mantle is made of (olivine, pyroxene, garnet) because their atoms are too big or are chemically incompatible - elements like rubidium, uranium, lead and strontium. These elements tend to rush into the melt if the rock starts to melt, so they are preferentially concentrated in erupted magmas and *depleted* from the residue remaining. This explains why they have higher concentrations in MORBs, relative to the mantle. However OIBs have even higher concentrations that are hard to explain by just a single melting event. So Hofmann and White proposed there had been *two* melting events (or more) in sequence, each of which would concentrate the incompatible elements.

Clem Chase, then at the University of Minnesota, independently reached a similar conclusion in 1981 just from the lead isotope data, like panel (c) of Fig. 10.3. Chase noted something that had been remarked on since the first measurements of lead in OIBs, by Paul Gast and others in the 1960s, namely that they plot mostly to the right of the so-called *geochron*, which is the line things would plot on if they derive from the original lead composition by a single melting event. This would imply negative or future ages: in effect, there would not have been time yet to accumulate enough ^{206}Pb to carry them to the right of the geochron. Geochemists are very clever (and I mean that) at measuring tiny quantities of exotic elements in rocks, but they are not so clever as to measure a rock that has not yet formed.

On the other hand, looking at the data in a different way, the *slope* of the array of data can also be interpreted as an age if there has been a second melting event that

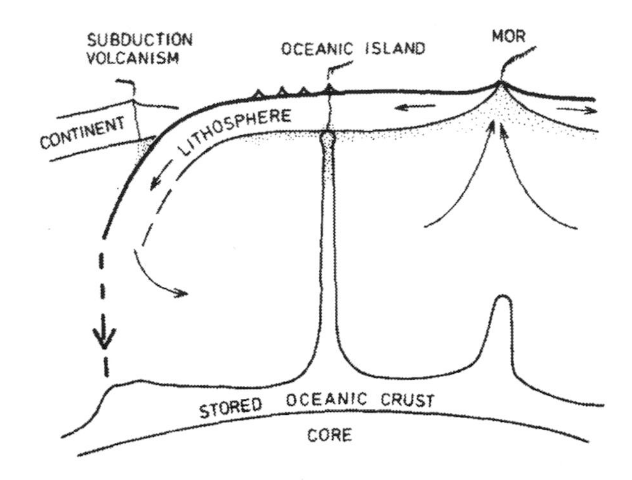

Fig. 10.5 Sketch by Hofmann and White of their proposal that OIBs derive from ancient MORBs that have been recycled through the mantle by subduction, sinking to the bottom, and being carried up again in a mantle plume. From [26], reproduced with permission

boosted the uranium content enough to generate the 'excess' ^{206}Pb. That age would be about 1.8 billion years. That's interesting, because strontium isotopes had also suggested a similar age. In both cases it was not clear if this might be a real age or a spurious alignment cause by other processes, so it was difficult to know what to make of the data.

Chase went further, and pointed out that the lead data for individual island groups each defined an apparent age, and the ages ranged from nearly 3 billion years down to about 1 billion years. This suggested each island group was revealing its own ancient melting event. There was a coherence in the data, which we won't go into here, that was unlikely to be due to chance, but that made sense if the data recorded ancient melting events. Chase argued that the ages are real, and that the sources of OIBs do indeed record a previous, ancient melting event.

Hofmann and White, meanwhile, argued that the incompatible-element enrichments could be explained if the ancient melting event was the formation of a previous generation of mid-ocean ridge basalt, in other words if the event was melting under an ancient mid-ocean ridge. Chase was not quite so explicit but implied a similar interpretation.

Hofmann and White proposed the following sequence. Ancient MORB was formed, carried along at the surface, then subducted. It is thought to be denser than average mantle, so it would tend to sink to the bottom of the mantle. Because it had more radioactivity than average mantle, it would get hotter and would eventually rise in a mantle plume. As the plume approached the surface it would melt and produce ocean-island basalt. OIB would not be the same as MORB because its plume source would not be normal mantle. Their sketch of the system is shown in Fig. 10.5.

The accepted later version of this sequence differs only in that a plume forms because material at the bottom of the mantle is heated by heat conducting out of the core, rather than by its internal radioactivity. Hofmann and White noted that seismologists had for decades found that the lowest 100–200 km of the mantle seems to have anomalous seismic wave velocities, and that their proposed accumulation of subducted oceanic crust could explain that layer which, for historical reasons is known as the D" (D-double-prime) layer.

There is no primitive mantle reservoir in Fig. 10.5. There are two reservoirs, the stored oceanic crust and the rest of the mantle. Later interpretations would say that there is a *greater concentration* of ancient crust stored at the bottom of the mantle, so there need not even be a sharp boundary between these 'reservoirs'. This seemed to be a very satisfactory explanation for the so-called *refractory incompatible elements* and their isotopes, but it was not clear how the noble gases could be accommodated. So the arguments raged on.

Hofmann, White and Chase not only took a fresh approach to interpreting the isotope data, they took a refreshingly low-key approach to arguing their cases. Al Hofmann is always the gentleman in my experience, focussing on the issues and open to a range of views, though I might wish he read my papers a little more closely. Bill White has a slightly gruff manner but I always found him friendly and he was one of the few geochemists to say simply that he liked some of my ideas. Clem Chase is an easy-going person who produced important insights on a variety of topics, one of the few to straddle geophysical and geochemical topics; we may meet him again in these stories. So there was a refreshing lack of egos.

Late in 1982 I was due for a semester without teaching at Washington University St. Louis and I determined to use the time to have a more comprehensive go at fitting the geochemistry into the physical picture of the mantle. I submitted the result, a longish paper, in January 1983. On checking now I see that a revision was submitted in October 1983 and it was accepted for publication exactly a year after first submission, in January 1984. That is how long it can take for reviews of a contentious paper to be forthcoming (nine months!), for the paper to be revised accordingly, for a second round of reviews, for arguing with an Editor and for the paper to be accepted. Then it was another six months, July 1984, before it saw the light of day.

So the process took almost two years, which was annoying, but it is not an uncommon experience. In the meantime I had moved back to Australia, and delays like that were to be more consequential.

My 1984 paper attempted a synthesis of the physics and chemistry of the mantle. It was called an 'interim synthesis' because the subject was developing so rapidly, but it was an attempt to promote the idea of looking at *all* of the evidence, geophysical and geochemical.

A distinctive feature of this paper was to consider the three isotopic systems, neodymium, strontium and lead, in a consistent way. An immediate implication was that the mantle is heterogeneous, because the lead and strontium together require multiple source types, as had been recognised for some time by the relevant geochemists, such as Paul Gast and M. Tatsumoto. There is also no obvious indication of a primitive source type. As there did not seem to have been a mass balance done for lead, I presented one (possibly for the first time), with implications for the age of the Earth as well. Others had tried to accommodate some heterogeneity by modifying the two-layer model, but to do that is to remove the original motivation for the two layers.

I'm interested now to see that the MORB and OIB isotopic data were characterised as being comparable: the diverse trends in the OIB data are also present in the MORB data, but with a smaller spread. I had forgotten I came to that view so early. It was a view that was soon to be stressed also by Al Hofmann, who brought new data to bear.

The noble gas data of the time were addressed, with the conclusion that they demonstrate some parts of the mantle have been 'degassed' less than other parts, and some of the degassing was very early, at least 4.4 billion years ago. These are much less stringent inferences than that there must be a large, 'undegassed', primitive reservoir. On the other hand I did not seem to address the missing argon question directly.

Noting more evidence that plate-related flow penetrates into the lower mantle, the paper goes on to advocate an un-layered, heterogeneous but possibly somewhat stratified mantle, the idea being that heterogeneities might be more concentrated in the lower half of the mantle. This might be due to an increase in viscosity with depth, for which there was some evidence, or to a tendency of heterogeneities to sink because they are denser than surrounding mantle. The latter would be a less efficient version of the sinking and separation proposed by Hofmann and White.

It is striking to me now that there is little discussion of plumes. The paper is focussed on distinguishing between the uniform-layers conception and the dispersed-heterogeneities conception. A few years later I determined the heat flow carried by plumes much more definitely, and that led to more explicit attention to plumes and their source.

The paper was in several aspects indeed an 'interim' synthesis, but it discussed much of the evidence and debate of the time in some detail. The treatment of lead was extensive. One interesting result was an age estimate of 4.52 billion years for what was probably the separation of the metallic core from the mantle; this was a bit older than previous estimates, and more consistent with estimates that the growth of the Earth took around 50 million years. It was striking then and now that there were almost separate communities, one dealing with lead (and strontium) and the other dealing with neodymium (and strontium). I got to know the 'lead' community a little, over time, including on some field trips to ancient crust. People like Wasserburg and Allègre certainly knew about the lead work, so it is curious it seemed to be blinkered out of their thinking.

That paper got some moderate notice in the business, though it did not seem to greatly influence the main debates. Perhaps more importantly it continued to define and frame my own developing work, and that led to some more definitive work over the next few years.

Also late in 1982 I had been alerted by a friend that a position was coming available at the Australian National University. This was the institution that most appealed to me, as it had an Institute of Advanced Studies that allowed for appointments to do mainly research, with only small teaching obligations. I did not mind the teaching, and it had expanded my knowledge and other aspects of my work, but I would be happy to have less of it to do. The other Australian universities tended to have quite heavy teaching loads, so they appealed less and had not been attractive enough to draw me back home.

The ANU was certainly an attractive prospect. It was one of the leading research universities in the world, despite its youth. There would be colleagues doing a range of relevant and exciting work. Australia had (and still has) a habit of not celebrating its high achievers (except in sport), but the ANU was a place of unapologetic excellence. I aspired to join that and play my part in continuing its excellent work. I needed no persuasion to apply.

On the other hand I had been in the US for fourteen years and my life was pretty strongly embedded there. In fact I had begun to assume that I was probably in the US for good. Two American wives (not at the same time), two American daughters, and a lot of water had passed under the bridge, on both sides of the Pacific. Anyway my application succeeded, and we (me, wife number two, daughter number two) moved there in July 1983. The coming years would be a mixed experience.

Chapter 11
Making It a Science?

Australia had changed a great deal in the fifteen years I had been away. The long, rather somnolent reign of Prime Minister Menzies and his mediocre successors had finally ended and the country had been dramatically stirred up by the accession of the Whitlam government in 1972. This generated some strident politics, but it also initiated substantial cultural shifts. Though the Whitlam government was dismissed in dubious circumstances by the Queen's representative (the Governor General) in 1975, the cultural changes rolled on, and I found it an exciting place to return to.

In the arts there was a great deal of experimentation and creativity, drawing on a wide range of influences from Asia as well as the traditional European sources. It was said that Australia had finally outgrown the *cultural cringe.* That term had been coined in the nineteen fifties to describe how we always deferred to the fashions and judgements of London or New York, and did not have the confidence to assert our own values. It was a colonial attitude, a hangover from being founded literally as a collection of British colonies.

It was certainly true that in quite a few areas of the arts there was a new assertiveness. Growing up, I had seen only a handful of movies set in Australia, and they had been clichéd stories. It was only in the late 1970s that real Australian movies were being made again, after a fifty-year hiatus, and I started to see my own culture and country on the big screen, instead of British and American cultures. I wanted to be part of that cultural renaissance. The ANU had been founded in 1948 as an elite institution to establish world-class research and postgraduate education, and it had succeeded in many areas.

More broadly, I could see Australia was maturing, but it had not outgrown the cultural cringe as much as many liked to think. Having lived in the US, I had seen how Americans just went ahead and did the things they wanted to do. I might not always like what they did, but I could see them going their own way. They had their Revolution and explicitly threw off the colonial reins. We had never had that break from the mother country, and the apron strings were still there. Our media always wanted to know what the rest of the world thought of something we did, or might do. We rarely did anything that had not been tried somewhere else already. This was

G. F. Davies, *Stories from the Deep Earth*,
https://doi.org/10.1007/978-3-030-91359-5_11

especially true in politics, and still is. We used to take our lead from Britain. Then we were taking our lead from the US, and that has only got worse. The Whitlam government was upsetting the old orders and threatening their power. That is the worst sin. Whitlam's removal for highly questionable reasons served the interests of many, but not of the Australians who elected him.

I had quite a lot invested in my return to Australia. I wanted to continue figuring out how the Earth works and I wanted to be part of a distinctive culture that was expressing itself again, as it had confidently done early in the twentieth century. Nearly forty years later again, the country is slowly waking to the presence of a unique and uniquely ancient culture in our midst, one that our settler forebears tried to wipe out. We can hope to forge a new arrangement, one that honours and respects that ancient culture and that draws on the strengths of both indigenous and settler ways.

That prospect is precarious at the moment, as destructive political forces repressively enforce an old order that no longer serves us. They threaten even the fundamental notion of knowledge, and universities as holders of knowledge. The institution I retired from a decade ago struggles to maintain its early, brilliant success. The struggle between those who become fixed on certain ideas and those who wish to promote new perspectives continues, more intensely. Both external and internal vested interests were at work in my time at ANU.

I set about two main threads of research, and some of my colleagues soon developed a third important theme. My first thread was to explore quantitatively how slow, whole-mantle convection would interact with chemical heterogeneities, to see if it was viable for heterogeneities to remain coherent on billion-year timescales. The second theme was the long-deferred effort to make viscosity depend on temperature and see how that affected the flow associated with the plates. The third theme arose when colleagues did experiments on plumes that greatly clarified their role.

Mike Gurnis had joined me as a graduate student in my last year at Washington University. I had let him know I might be moving and he was keen to come anyway. When I did move he applied to join ANU and was accepted, so he was able to continue on our work.

Mike set about exploring how plate-related convection would stir passive tracers that float around in the flow, simulating the chemical tracers measured by geochemists. Mike's convection models basically picked up where Rick Lux's had left off. This was partly dictated by the available computer power of the time and also to keep that part of the work straightforward so he could focus on the stirring. It emerged that the stirring was not too sensitive to that choice. The main models simulated the subduction of heterogeneities and their later removal at a mid-ocean rise.

There are some subtleties about stirring and mixing heterogeneities in a fluid. One is to distinguish *stirring*, which is just the intermingling of two kinds of fluid, from *mixing*, which is taken to mean the fluids blend and become one. In the mantle the different rock types would maintain their chemical identity pretty much unless they were melted, so we argued the issue in the mantle is stirring, not mixing. Figure 11.1 shows an example of this kind of stirring.

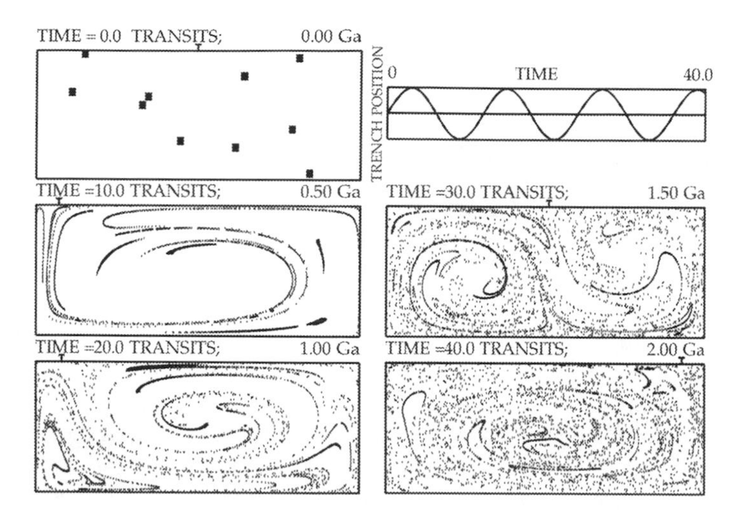

Fig. 11.1 An illustration of stirring of tracers in a flow that slowly changes from one cell to two and back again. The tracers were introduced in random clumps at the beginning and even after many transit times some clumps still persist. The total time would correspond to 2 billion years or more in the mantle. From Davies [4], reproduced with permission

Another subtlety is that flow that changes only slowly stirs much less efficiently than quickly changing flow. The slow shifting about of spreading centres and subduction zones means the plate-related flow changes fairly slowly. The flow in Fig. 11.1 is of this kind.

It makes a big difference if there are smaller-scale turbulent eddies in the flow. The flow in Fig. 11.1 has only the large-scale flow. A familiar example will illustrate the difference. If you pour some cream gently into you coffee, then a few strokes of a spoon will stir the cream (and mix it) into the coffee. On the other hand if you were to pour some cream into a jar of honey and stir that you would find you have to beat it for quite a while before you got the cream to blend with the honey. The reason is that the greater stiffness (i.e. the higher viscosity) of the honey suppresses any small eddies, whereas a spoon passing through your coffee spins off large and small eddies that greatly accelerate the stirring. The mantle is much more like a honey pot than a cup of coffee.

Mike's basic result was that some clumps of tracers could persist for quite a long time. In other words heterogeneities had quite a long *survival time*, which is the time for which a heterogeneity is still identifiable. Scaled to the whole mantle they would last a billion years or more. This was much longer than others had been estimating based on upper mantle models with more complicated 'small-scale' flows, which tended to show heterogeneities homogenised with a few hundred million years.

The *residence time*, on the other hand, is the average time between a tracer entering the flow and being removed at a spreading centre. Scaled to the whole mantle the models yielded between 1 and 2.3 billion years, bracketing the age of about 1.8 billion years inferred from lead isotopes. If the models were scaled to the upper mantle the residence times were only 400–500 million years. These results encouraged our hypothesis that the isotopic heterogeneities could be consistent with whole-mantle convection.

Mike explored the effects of the mantle viscosity being greater at depth, and of tracers that are heavier than surrounding fluid. The latter had only modest and temporary effects in these models, though later work would show greater effects. On the other hand if the deep mantle is up to 100 times more viscous than the shallow mantle then survival times can be increased, so that some material might survive from quite early in Earth history.

These studies suggested that whole-mantle convection might indeed be consistent with observations of isotopic heterogeneities, but they were still rather simplified. Others were reaching different conclusions, mainly because, we argued, their models were not well matched to the mantle. Dan McKenzie and his student N. Hoffman, for example, found heterogeneities were rapidly dispersed or destroyed in upper mantle cellular convection. Don Turcotte and his student Louise Kellogg considered only the kind of flow that maximises stretching and thinning of heterogeneities and found that they are removed efficiently. In a follow-up paper, Mike showed that some small heterogeneities plausibly persisted for much longer, even as such stretching and thinning proceeded for other heterogeneities.

Uli Christensen had already published some innovative convection models featuring more complicated viscosity laws. In 1985 he presented a careful series of models that further clarified the apparently conflicting results on stirring. There are flows that stir quite efficiently, there are other flows that stir inefficiently, and there are flows that leave parts of the fluid free of heterogeneities. Generally the efficiently-stirring flows are more time dependent, and particularly tend to have cells that break up and reform.

In a published comment I complimented his instructive work but questioned his conclusion that the mantle would be well stirred. In his reply he acknowledged my points but argued that he still thought the mantle would stir efficiently, because even a little of the wrong kind of unsteadiness could have a big effect. So we agreed there are subtleties of stirring and disagreed in our assessments of what might be happening in the mantle. Our civil and constructive exchange was in contrast to the attitudes of McKenzie and Turcotte, who paid less attention to the subtleties of stirring and tended to disdain whole-mantle convection when they did not ignore it altogether.

Our understanding of these issues would shift significantly as the form of mantle convection was clarified, and by later work on stirring.

Mantle convection modelling had a bad reputation. People outside the field tended to see a diversity of very simplified models that did not look much like the Earth, or each other. They saw practitioners squabbling. Most importantly, there seemed to be few constraints that could be brought to bear to choose among the models, and to test whether the models had any useful resemblance to the Earth. It was not uncommonly described as ‘computer games’. That was not quite fair, as the subject was in a phase of exploration, of figuring out which factors might be the most important and how such models worked. Also the early models were very limited by available computer power. But many of the models, in my view, were not well conceived to teach us much. The 'computer games' accusation stung, but there was more than a grain of truth in it.

I was becoming more convinced that strong observational constraints *could* be brought to bear, but to do that I needed the models to incorporate a pretty good simulation of a plate. That would allow reasonable comparisons with observations of heat flow, topography, gravity and the geoid. I recall a conversation with a visiting mineralogist who asked me what I did. I described how I was modelling mantle convection so I could compare the models with well-established observations. He did not seem to hear me, or believe what I said was possible, because he ended the conversation by saying he preferred to work on things one could measure. It is of course the essence of science to compare your hypothesising with what can actually be observed. I wanted to make mantle convection a science.

The most important thing still missing from my convection models was something resembling a stiff plate. The reason the lithosphere is stiffer than the underlying mantle is that it is cooler. The slow, creeping deformation that can occur at mantle temperatures of around 1300 °C rapidly slows to a negligible amount with a temperature drop of a few hundred degrees. A piece of the cool lithosphere, a plate, then moves as a coherent unit. Our previous modelling had shown that even just imposing a uniform velocity on the surface of the fluid induced more plate-like behaviour, as in Figs. 9.5 and 9.6. However the model in Fig. 9.6 has a constant viscosity, which leaves the cool fluid at the top freer to drip off and fall through the layer. Models by others, such as Mark Parmentier and Uli Christensen, had shown that making the cool fluid more viscous tends to suppress these drips and make the behaviour more plate-like, but their models had other complications that might or might not be appropriate for the mantle.

I was keen to do some models that I considered would be more carefully tailored to the mantle, particularly the whole-mantle case. No-one else seemed to see the need for stiff plates as strongly as I did. Although others had done models that encompassed some plate-like behaviour, they had not looked systematically at how to include them, what their effects are and how well the models then compare with key observations.

I knew it might well be challenging to take the next step and include variable viscosity, but I thought anything less was a waste of time. So here I was, at a top institution, with modest teaching requirements, reasonably good computing facilities for the time, and an aspiration to make contributions worthy of the high reputation of ANU. It was time to be adventurous and take on a riskier project than I had had time for so far. It seemed the straightforward thing to do, to be ambitious and not waste the opportunities I now had. I set about doing models with temperature-dependent viscosity.

It was easy enough to prescribe the fluid viscosity to be higher at lower temperatures. It was not so easy to get the computer then to solve the convection equations. With constant viscosity the flow equation could be solved with a fairly direct and routine procedure. With viscosity varying from place to place in the fluid that method did not work. In such cases you can modify the procedure to start with an approximate solution and then solve for corrections that improve the solution's fit to the governing equations. In favourable cases a few passes of correcting can get you a quite accurate solution. The mantle case is not very favourable. The reason is that the viscosity varies by about a factor of 10 for each 100 °C change in temperature. The lithosphere would be many factors of 10 more viscous than the interior fluid. In that situation the 'corrections' to an approximate solution can make it worse. The approximate solution 'blows up' instead of converging to an accurate solution.

I set about finding a method that would be more robust, so it would yield a solution reasonably reliably and without taking too long on the computer. That search turned into a saga. A few others had used a fundamentally different approach, called finite elements, whereas I was using finite differences. Even if they had readily shared their methods, it would have been a major re-tooling for me, and finite element calculations are slower. It would be better if I found a satisfactory fix for my finite-difference approach, so I pursued what seemed to be a path of least resistance. Others had used finite differences, and I was already using the method used by Houston and De Bremaecker, with some refinements, but I needed to work at finer resolution, and that slowed the convergence. There were better methods out there, but it took me a long time to find them.

I searched the relevant mathematics literature, and there were many claims for methods that might work, but their problems were generally not as extreme as the mantle case. I spent quite a lot of time trying methods claimed to be robust but found them to fall short of the claims. I discovered there is quite a lot of rather mediocre work in other disciplines, where rather inflated claims were made for not very dramatic advances. I interacted with computational people across the campus, but few had my kind of problem to deal with, and there was often a language problem—some would fire questions at me in jargon I was not familiar with, and lose interest if I did not seem to know what I was talking about (in their language). It would have required a careful joint exploration for them to understand my needs and for me to understand their methods. Some tried to be helpful but did not seem to have what I needed. You get the idea.

I began to find methods that helped. There are various things, some simple and some sophisticated, that can improve the robustness of the computation. A very simple one is to only add a fraction of the new correction, and it often keeps the procedure stable. Another simple one is to limit the range of viscosity. You don't need more than a factor of about 100 variation, because that is enough to make the lithosphere stiff enough to behave like a plate. I reformulated the calculations in ways that seemed more favourable. I found more sophisticated things like a 'conjugate gradient' method that would accelerate convergence.

Eventually I was satisfied that I had models of sufficient reliability and resolution to be useful. For the sake of the history I will show the model that first seemed to accomplish what I wanted, in Fig. 11.2, as published in 1988. The height of graphics technology at the time was a computer-driven pen plotter. There were no laser graphics or grey-scales. So you got line plots, with the blocky letters and numbers that came with the plotter. The essence can be seen in the 'T' (temperature) plot: a cool, stiff plate extends the length of the box, forming part of a single convection cell extending the length of the box.

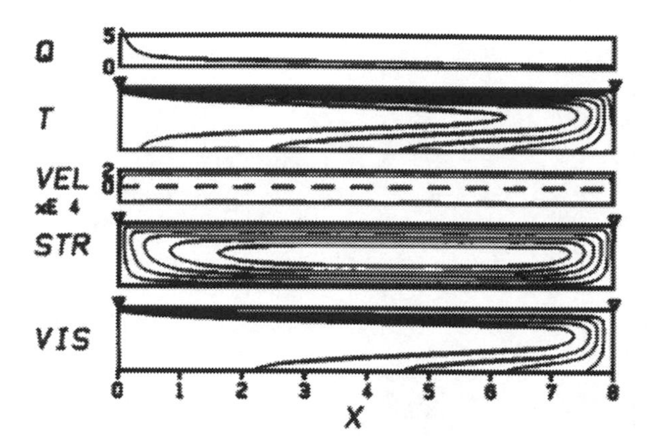

Fig. 11.2 A convection model with a stiff upper plate moving at an imposed surface velocity close to the velocity the unrestrained fluid would move at. **Q** denotes surface heat flux, **T** shows isotherms, **VEL** is the surface velocity, **STR** shows flow lines (known as streamlines in the business), and **VIS** shows contours of viscosity. There is a single convection cell extending the length of the box, with the plate being the upper 'thermal boundary layer'. From Davies [27], reproduced with permission

A model that is clearer and may be more convincing is shown in Fig. 11.3. It is from a decade or so later using a better computation method, a more powerful computer and rather nicer graphics. It shows something like a plate moving under its own impetus, with no imposed velocity. Rather, the surface is 'free slip', as if it were only air above the fluid surface. Flow lines are plotted over the temperature on the left, and viscosity is shown on the right. The viscosity depends only on temperature, except in two places where it is reduced to the minimum, at the left end of the box and towards the right end. The effect of these breaks is to free up the left-hand segment of the top cool layer so it can move easily, whereas the right-hand segment is effectively tied to the end of the box. The result is that the left-hand segment moves to the right and sinks at the viscosity break, very much as though it were a plate subducting. The sequence is continued for some time so you can see how the plate sinks to the bottom, folds and spreads along the bottom. Meanwhile new plate forms at the left-hand end where the flow rises to the surface.

I had in 1989, soon after publishing Fig. 11.2, achieved a model in which the plate moved under the action of the negative buoyancy of the sinking plate, but the model in Fig. 11.3 is clearer, so that's what I'm showing here. I will refer back to a number of features of this model, but for now I want to highlight that the moving plate has a nearly constant velocity along its length, which you can tell from the outer flow line being parallel to the surface. On the other hand the other plate does not move much at all, as the wider spacing of the flow lines indicates slower velocity. Both plates stretch a little bit, so they are not as 'rigid' as real plates, but they are quite a good simulation. The result of of these motions is that the 'subduction' is fairly asymmetric, meaning only the left plate goes down. This is a key feature of real subduction zones that accounts for the oblique descent of the subducted 'slab'. This was foreshadowed by

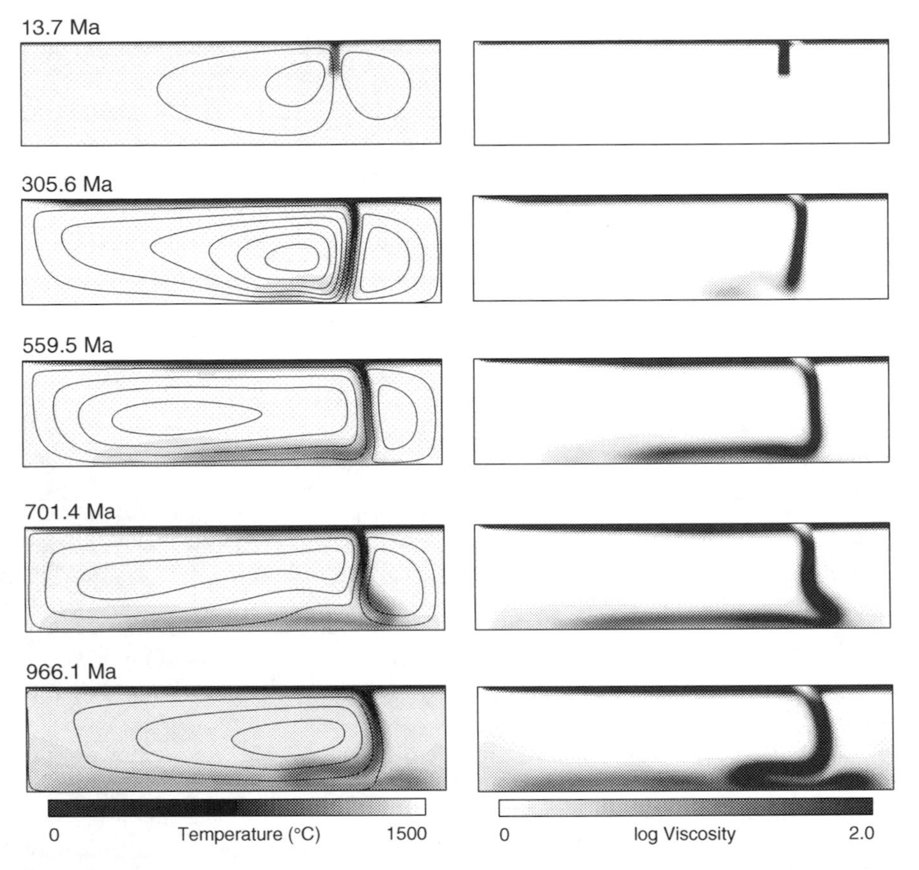

Fig. 11.3 A sequence from a convection model in which the viscosity depends on temperature, except for two breaks of low viscosity at the left end and near the right end. The left-hand segment of the cool top layer is thus free to move. It was started with a small, cool piece projecting down, in order to get it moving to the right. It behaves much like a plate forming and moving at the surface, subducting and merging into the deep fluid. *Ma* means million years. From Davies [4], reproduced with permission

the much simpler model of Fig. 9.5, that includes only an imposed moving boundary, and by Brad Hager's equivalent models in three dimensions.

The box in Fig. 11.2 is very long, eight times the depth, because I wanted the effect of the plate to be clear, and because later models had two plates with a spreading centre in the middle. It was possible to calculate observable quantities from this model, notably the surface heat flow and the surface topography. These quantities relate to two additional striking features that had emerged from exploration of the sea floor.

We need to take a look at some more observations. The mid-ocean rise system is not just the second-largest topographic feature on the planet (Fig. 11.4), it was

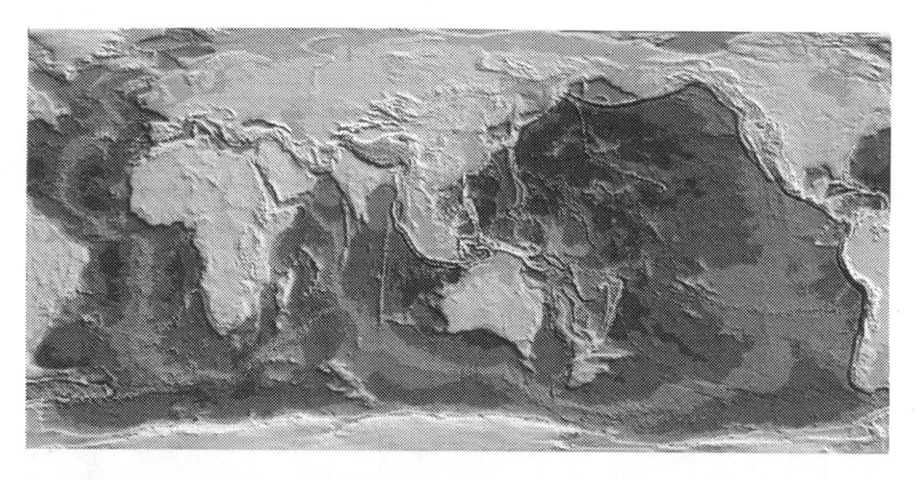

Fig. 11.4 Topography of the Earth. The submarine breaks in the grey scale are at depths of 5400 m, 4200 m, 2000 m, and 0 m. Shading of relief is superimposed, with a simulated illumination from the north-east. From the ETOPO5 data set from the U.S. National Geophysical Data Center [28]. Image generated using *2DMap* software, courtesy of Jean Braun, then of Australian National University. From Davies [4], reproduced with permission

recognised as being related to the seafloor spreading occurring at its crest. The elevation of the rise system was taken to be an expression of upwelling convection. Yet a curious regularity was found in the rise topography.

John Sclater, R. N. Anderson and M. L. Bell found in 1971 that if seafloor depth is plotted *versus* the age of the seafloor, much of it falls close to a single subsidence curve. This is particularly true for sea floor younger than about 70 million years; after that the depth is less regular. Even more remarkably, the subsidence closely follows a regular mathematical form: it is proportional to the square-root of the age. In other words sea floor of age 64 (square-root 8) million years has subsided twice as much as seafloor of age 16 million years (square-root 4), which has subsided twice as much as seafloor of age 4 million years (square-root 2), and so on. This is shown in Fig. 11.5.

This particular form has a known physical explanation. It is what you get if the plate cools by heat conducting to the surface as the plate drifts away from the rise crest. This was demonstrated most clearly by E. E. Davis and Clive Lister in 1974.

The same explanation predicts how the surface heat flow will vary with time as the plate cools: the heat flux varies in *inverse* proportion to the square-root of time. Thus at times 4, 16 and 64 million years the heat flux will be half, one quarter and one eighth what it was at 1 million years. Figure 11.6 shows the measured heat fluxes coming through seafloor of various ages, compared with the prediction (the 'simple cooling model'), which it fits remarkably well.

By the way, *heat flux* is the amount of heat that flows through one square metre in one second. *Heat flow* is the total amount of heat from some larger area.

There is another curve in Fig. 11.6, called 'plate model', because some of the early plots of depth *versus* age suggested that the sea floor reaches a constant depth

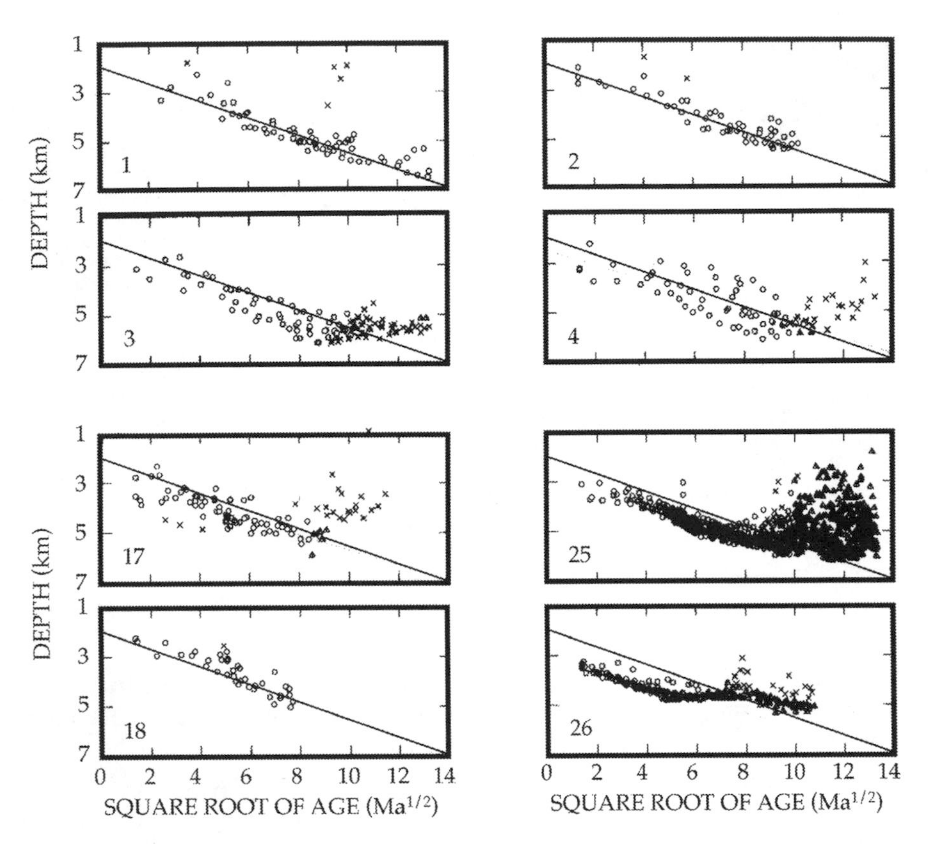

Fig. 11.5 Seafloor depth *versus* square-root of age for a selection of regions of the seafloor. The same reference line is shown in each plot. After Marty and Cazenave [29], reproduced with permission

after about 70 million years; it was said that the sea floor 'flattens'. I have shown a selection of regional plots in Fig. 11.5 to show that there is quite a lot of variation of older sea floor: some continues to subside, some rises again, some levels out or rises then sinks again. This clearly indicates that other things are happening than just the lithosphere steadily cooling. We can potentially learn useful things from that.

However another story had taken hold based on the early data. The story is that plates reach a maximum, steady thickness and this causes them to stop subsiding. You could maintain this story so long as all the seafloor data were aggregated into a single plot: the scatter increases but the data are not inconsistent with flattening. However J. C. Marty and Annie Cazenave published their regional plots in 1989, of which Fig. 11.5 includes a selection, and they show that there is no simple flattening: it is an artefact of aggregating all the data into one plot. Nevertheless the (constant-thickness) 'plate model' became a small industry. There may still be papers being published purporting to explain seafloor flattening. It is yet another illustration of

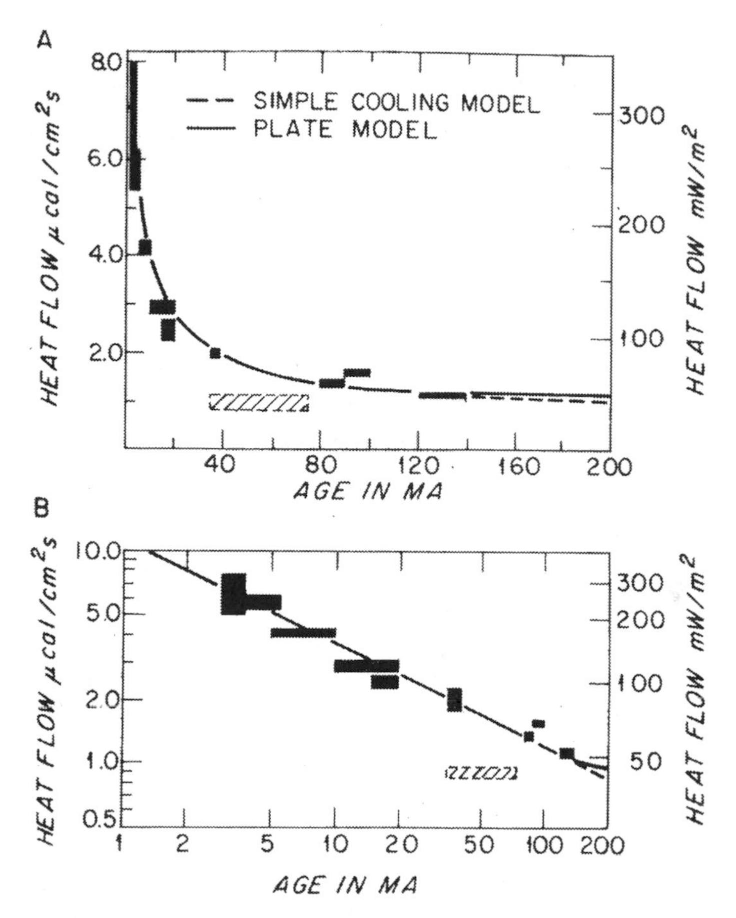

Fig. 11.6 **a** Oceanic heat flux *versus* age of seafloor. The two curves are explained in the text. The shaded box represents observations that were superseded by the data at 40 Ma. **b** The same data and curves on a logarithmic plot, in which the 'simple cooling model' predicts a straight line with a slope of $-1/2$. From Sclater et al. [30], reproduced with permission

the coloniser effect, in which the first interpretation takes root and is very hard to displace.

One of the reasons the plate model continued to have a life was that it was supposedly consistent with the idea of small-scale convection cells in the upper mantle, championed by Dan McKenzie among others. I could explain, but won't here, that it depends on how heat is delivered to the putative small-scale convection as to whether you get a flattening, on average, or a steepening, and that issue was either not recognised or glossed over. There is also the question of whether the small-scale convection should have more directly observable effects, on the topography for example. We will come back to that.

We can now return to the question of *how plates relate to mantle convection*, and to the 1988 model in Fig. 11.2 and the paper in which it was presented, entitled

Role of the lithosphere in mantle convection. The paper first makes a fundamental point. Convection is a heat transport mechanism, a way of transferring heat through a fluid layer. In the case of the Earth, the heat lost from the interior to the surface is dominated by the heat flowing through the sea floor (Fig. 11.6), and that is controlled by the plates, as the regularity of the heat flow reflects the slow cooling and thickening of a plate. There is a cycle of mantle rising at a spreading centre, drifting horizontally at the surface and losing heat as it goes, then subducting and sinking back into the mantle. The net result of this cycle is to remove heat from the mantle, in other words to transport heat. The numerical models, and some simple estimates made later, show that the motion of the plates is maintained by their negative buoyancy as they subduct. Thus this cycle is a form of convection. Furthermore, the plates are an *integral part* of the convection, in fact the main active component.

Whereas others had been conceiving of convection *under* the plates, whether the large cells suggested by Holmes and Hess or the smaller, upper mantle cells often discussed at the time, this argument said the plates are the most active part of the convection, and the convection comes right to the Earth's surface. Wilson's description of plates and plate boundaries is a description of a brittle solid, and that accounts for their odd shapes and sizes, yet the plates are still an active part of a convection system.

The paper then presents quantities calculated from the convection model: heat flow, topography, geoid and gravity perturbations. I don't need to show them here, the heat flow fits the observations in Fig. 11.6 quite well. The calculated topography has a form close to that in Fig. 11.5. The geoid and gravity variations are of lower amplitude than observed. It is demonstrated that the geoid is very sensitive to details, being a small difference between large quantities, so it is not to be expected to fit well in these first approaches to modelling, and the gravity is deduced from the geoid. The topography is also of a little smaller amplitude than the simple cooling model predicts using the model parameters, but we don't actually know the material properties of the real oceanic plates that well, so again a small misfit is within uncertainties and not a serious issue. The key result is that these observables have the right form and are reasonably close to observed amplitudes.

For all the many computations purporting to model mantle convection, no-one had systematically compared model results with observations in this way. The many models without any lithosphere could not remotely approach the forms in Figs. 11.5 and 11.6. To achieve first-order consistency with observations was a significant step in the right direction. It might be still have been a bit approximate, but this was not just computer games, it was science.

The next step was to exploit the approach to learn more about the mantle. I did that in two following papers, entitled *Ocean bathymetry and mantle convection*, Part 1 *Large-scale flow and hotspots*, Part 2 *Small-scale flow*. *Bathymetry*, a term used by oceanographers, is the measurement of depth below the sea surface, so effectively these papers are about seafloor topography and mantle convection. They make the point that seafloor topography is strongly affected by mantle convection, and is thus a primary constraint on models.

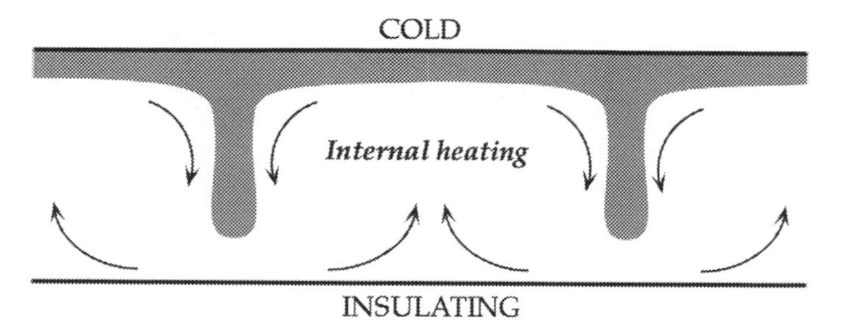

Fig. 11.7 Sketch of convection when there is internal heating and no bottom heating. There are no hot active upwellings, only cool downwellings

In *Part 1* I presented a series of computed models and also laid out a key insight, a general relationship between topography and heat flow. This enabled a test of two-layer convection versus whole-mantle convection. It also yielded a key observational measure of plumes, and showed how to avoid a conceptual stumbling block that had bothered the subject since the early days of seafloor exploration.

The explanation of how heat flux and topography vary with plate age (Figs. 11.5 and 11.6) involve an implicit assumption, namely that the mantle beneath the cooling plate is passive. Evidently this is true to a first approximation, or the relationships would not be evident. Yet there are deviations in the topography, more evident with older sea floor, and also evident in variations in the depth to the crest of the mid-ocean rises, which vary by about a kilometre. The latter can be seen in Fig. 11.4 where, for example, the crest south of Australia is quite deep but near South America it is broader and shallower. As rise crests are of zero age, the latter variation cannot be due to differential cooling and must be due to the influence of underlying mantle. So some moderate influence of the underlying mantle is evident.

This logic had led me to the view that the mantle is mainly heated from within, rather than from below. Since the work of Arthur Holmes, and later Harold Urey in the 1950s, the mantle was believed to contain enough radioactivity to roughly balance the heat escaping through the Earth's surface. Figure 11.7 is a sketch of convection when there is only internal heating (it can be compared with Fig. 9.1, which is for normal Rayleigh-Bénard convection with both heating at the bottom and cooling at the top). With no heat coming from below, there is no hot thermal boundary layer at the bottom, and consequently no hot upwellings. With cooling at the top, there are cold downwellings that drive the flow. Fluid away from the active downwellings is passively displaced sideways and upwards.

The model in Fig. 11.2 is of this type, with only internal heating. *Part 1* of the next papers presented four computed models, two internally heated and two bottom heated. Each had two plates separating at a spreading centre. In the first of each pair the spreading centre was stationary, and in the second of each the spreading centre was migrating slowly according to the rules of plate movements outlined in Chap. 8.

The first of these four models was similar to Fig. 11.2, just with two smaller plates. When the spreading centre was allowed to migrate the topography became

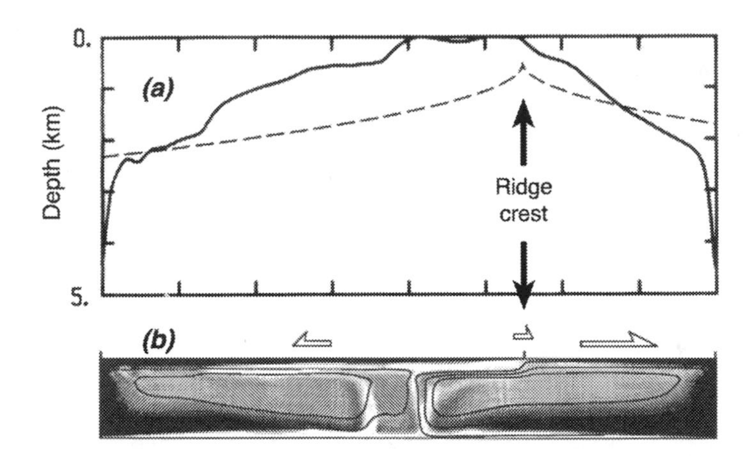

Fig. 11.8 **a** Topography (solid curve) from a model with bottom heating and a spreading centre that has migrated to the right, starting in the middle. The dashed curve is the cooling lithosphere prediction. **b** Temperature (grey scale) and flow lines. Arrows above indicated the velocities of the plates and the spreading centre. From Davies [31], reproduced with permission

somewhat asymmetric and did not fit the standard subsidence model as well, yet both plates did subside. The reason for the asymmetry was that the deep displaced fluid still drifted towards the centre of the box before ascending, and being somewhat warmer it raised the surface a little on the side of the longer plate. It is to be expected that the deeper fluid will have some influence on the topography, because it does not become instantly homogenised, and remaining small internal temperature variations will have some influence. This would be part of the explanation for the non-plate topography of the real sea floor (Fig. 11.5).

Bottom-heated models yielded markedly different topography. An example is shown in Fig. 11.8, which has a migrating spreading centre. Several hot upwellings have developed because the bottom heating has generated hotter fluid at the bottom that has lower viscosity and is more unstable. The ridge has migrated to the right, having started at the centre, but the upwellings are still near the centre. The resulting topography has an extra kilometre or more of elevation of the left-hand plate. The geoid and gravity anomalies, not shown here, are equally irregular.

Now to the lessons about the Earth. If convection were confined to the upper mantle, it would have to be heated mainly from below because there is not enough radioactivity just in the upper mantle to account for the heat emerging through the sea floor. This means most of the heat emerging at the sea floor must be coming from deeper, from the lower mantle. Thus the upper mantle would have to be heated from below, there should then be a hot thermal boundary layer at the bottom of the upper mantle, and hot upwellings should rise from it and generate large topography. Seafloor topography should have large irregularities, more like Fig. 11.8, rather than the regular subsidence actually observed (Fig. 11.5). These results argue strongly against the upper mantle convecting as a separate layer.

On the other hand if the mantle convects as a single, deep layer then its heating should be mainly from within. There will be some heat coming from the core, but it will be secondary, as we will soon see. The results from the internally heated models should therefore be a good approximation to whole-mantle convection. The internally heated models generate topography similar to what is observed, so whole-mantle convection is supported.

The results of *Part 2* of *Ocean bathymetry* can be quickly summarised. Any form of so-called 'small-scale' upper mantle convection would generate substantial topography if it transported a significant amount of heat. One such mode is cells or rolls confined between the lithosphere and the bottom of the upper mantle at about 650 kms depth, as proposed by Frank Richter, Dan McKenzie and others. If such a mode transported the heat emerging from the older sea floor, as they propose, it ought to generate topography of 1–2 kms peak to trough. Nothing like that is evident on the sea floor.

The other possible small-scale mode is basically drips off the base of the lithosphere, the idea being that the lowest part of the lithosphere is the warmest part, and therefore the least viscous. It might therefore be prone to dripping away. If it did, at a rate that transported significant heat, it also should cause topography of about a kilometre, peak to trough. No such topography is evident. There were some small geoid anomalies in an appropriate pattern in a couple of restricted areas, but they are of such small amplitude as to involve only a very minor amount of heat transport.

A further point is that if such small-scale drips did occur they would thin the lithosphere, increase the heat lost through the lithosphere and cause the surface, on average, so subside *faster* than the simple cooling model. It can be seen in the model results, which I have not shown here. This effect is the opposite to the putative effect claimed: rather than causing the sea floor to 'flatten' it would cause it to *deepen*.

In the course of doing the work just summarised I realised there is a close relationship between topography and heat flow. It arises in a convecting fluid because both depend on buoyancy. I don't recall how I came to the idea, but probably through a habit of checking to make sure the results of numerical models make sense. This is an excellent habit to cultivate, as it is all too easy for a mistake to hide in computer code, and also because it can lead to a clearer understanding of what the model is doing. A computer model might yield a plausible-looking result, but if it is a complicated model the reason for the result might not be any more evident than the reason the Earth does what it does. *Understanding* is far more valuable that a single result.

The relationship between buoyancy and topography is obvious enough: the buoyancy that can cause the fluid to rise can also lift the surface. The basic idea is illustrated in Fig. 11.9. If the blob 'a' is buoyant it will lift the surface. It's our old friend the isostatic balance, from Chap. 5: the extra mass in the topography is balanced by the deficiency of mass in the blob (if it is buoyant its average density will be less than its surroundings). A blob that is near the bottom of this fluid layer, like 'b', will not have much effect on the top surface, but it will lift the bottom surface. A blob in the middle, like 'c', will have an effect on both the top and the bottom, though the topography will be low and broad and may not be easily detectable.

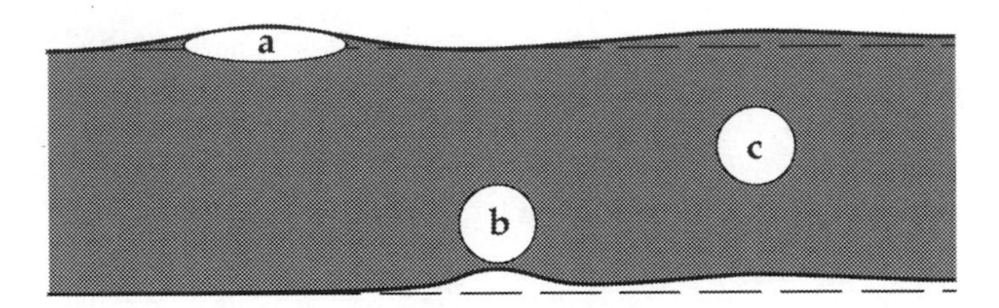

Fig. 11.9 Topography generated by buoyant blobs

This principle applies to the subsiding oceanic lithosphere as well. As the lithosphere cools it thermally contracts and its density increases. Its top surface subsides because of this contraction. The amount it subsides is determined by the isostatic balance: the amount of mass in a vertical column through old sea floor should be the same as in a column near the rise crest. The amount of mass has not changed, it has just shrunk a bit.

Now the reason the lithosphere has shrunk is that it has lost heat to the surface. The amount of heat lost (per unit area) bears a simple relation to the amount of contraction and subsidence. The old sea floor is about 3 kms lower than the rise crests. When you do the numbers for the amount of heat extracted to cause this amount of subsidence it matches the measured heat flow out of the sea floor (Fig. 11.6). As it should.

What if convection were confined to the upper mantle? Then, as I explained earlier, the upper mantle would be mostly heated from below, it would have a hot thermal boundary layer at its base, and that would generate buoyant rising columns. The buoyancy of that rising fluid would progress through stages like 'b', 'c' and 'a' in Fig. 11.9, so that when it reached the top it would lift the surface. How much topography would it generate? Well, if was carrying as much heat as is emerging out the top, then it would generate topography comparable to the mid-ocean rise system: something like the topography in Fig. 11.8. In other words upper-mantle convection should generate a second kind of topography comparable in scale to the topography of the mid-ocean rise system. Clearly there is no such topography of the sea floor. There are muted deviations from the broad subsidence of the cooling lithosphere (Fig. 11.5), but there is no broad, pervasive, anomalous topography 1–3 kms high.

The particular models presented in these papers make a strong case, but to that can be added a quite general argument. Upper mantle convection should generate large topography, but no such topography is evident. Therefore upper mantle convection is unlikely.

Well, in fact there *is* another kind of topography, though it is small compared with the mid-ocean rise system. Volcanic hotspots are typically surrounded by *hotspot swells* about a kilometre high and roughly a thousand kilometres across. The best known and clearest example is the Hawaiian swell, which is evident in Fig. 11.10.

Hotspot swells had been highlighted by the late Tom Crough in 1983. They were attributed to the effect of a mantle plume, as proposed by Jason Morgan. Figure 11.11 illustrates how a plume is envisaged to rise under the moving lithosphere and lift it. By looking at the cross-section CD, the balance between the topography and the

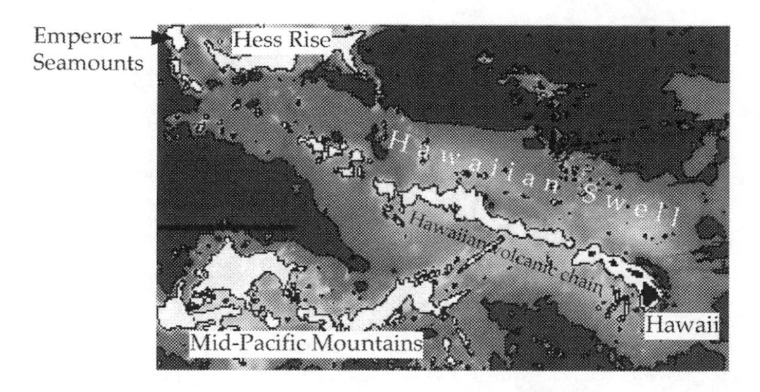

Fig. 11.10 The Hawaiian swell, surrounding the Hawaiian volcanic chain, which originates at the volcanic hotspot erupting at the 'big island' of Hawaii. The contours are at 3800 m and 5400 m. From Davies [4], reproduced with permission

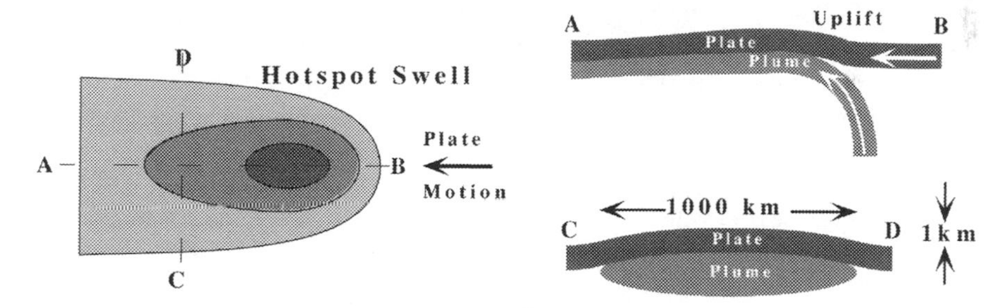

Fig. 11.11 How a plume generates a hotspot swell. The swell is sketched in map view on the left, commonly about a kilometre high and a thousand kilometres across. The section AB depicts a columnar buoyant plume rising under the moving plate, bending over, emplacing under the plate and lifting it. Section CD shows the relationship between the swell topography and the buoyant plume material below. From Davies [4], reproduced with permission

plume buoyancy can be calculated. The plate moves about 10 cm per year to the northwest, and in that time the equivalent of a strip of sea floor 10 cm wide and one thousand kilometres along section CD is lifted by about one kilometre. The weight of that new topography balances the buoyancy of the plume material that has arrived in that time. From the buoyancy you can calculate the heat being carried by the plume.

I won't bore you with the detailed numbers. The end result is that the Hawaiian plume carries about one percent of the total heat flow out of the Earth. If you go back to Fig. 11.4 you can see the Hawaiian swell in the mid-north Pacific. You can see it is distinct, but small compared with the mid-ocean rise system. There are a dozen or more distinct volcanic hotspots, each with a swell. I used Tom Crough's compilation to estimate that in total they carry between six and ten percent of Earth's heat flow. A couple of years later Norm Sleep of Stanford University did a more careful and complete assessment and came to a similar result. (It turned out Norm had foreshadowed the use of hotspot swells to estimate heat flow in a paper published around the same time as mine. We agreed we had come to the idea independently and were happy to cite each other. That joint recognition seemed to catch on, so both papers are commonly cited. Norm is a generous bloke and it is possible to be nice in science.)

The conclusion is that plumes carry no more than about ten percent of the Earth's heat flow. They are buoyant upwellings, but secondary in comparison with the cooling and subducting plates, which generate negatively buoyant downwellings. In my *Part 1* paper, discussed above, I used a couple of estimates of the heat likely to be flowing out of the core into the base of the mantle, and the answer is similar to the inferred plume heat flow. Thus the idea that plumes originate at the bottom of the mantle is nicely supported. Also supported is the idea of whole-mantle convection, heated mostly from within by radioactivity, but with some heat entering from below.

My ambition to include some reasonable semblance of a plate in convection models, by adding temperature dependence to the viscosity, had certainly borne fruit. The models yielded quantities that could be directly compared with observations, most notably heat flow and topography. That empirical testing strongly favoured whole-mantle convection over upper-mantle convection.

Just as important were the insights that emerged along the way. The most important is the general relationship between topography and heat flow, which allowed the plausibility of the model results to be checked and opened the way to quantitatively characterising mantle plumes. That in turn firmed the idea that plumes arise from the base of the mantle, driven by heat emerging from the core. It portrayed the plumes as a significant but secondary part of the convecting mantle.

Also, the instructive comparisons of models and estimates with observations meant it was science, not just theoretical exploration or computer games.

This all seemed very satisfactory. The considerable effort and frustration of getting the computations to work seemed to have been very much worthwhile. The middle paper, *Ocean bathymetry and mantle convection, Part 1, Large-scale flow and hotspots*, is currently my most cited paper. I suspect that part of its citation rate is because it is a default reference, along with Norm Sleep's, for the heat flow carried by plumes. Not everyone who cites it has necessarily read it and fully understood all its arguments and conclusions.

Why do I quibble about the 'success' of the paper? Because the other main conclusion, that the observed seafloor topography means upper-mantle convection is not viable, did not seem to have great influence. You might expect geochemists not to pay much attention, the concepts were not familiar to them, but the geophysical community was still divided. Don Turcotte, for example, continued to argue vociferously for layered convection, despite the physics of my argument being well within his competence to understand.

It is not uncommon for the first statement of a significant argument to be less clear and articulate than later versions. The papers are fairly long and complicated, and the computation methods were stretched near their limit. The model results were not as clear-cut as later models would yield. However I regularly expounded the argument against upper-mantle convection, at conferences and in papers, over the next decade or so. The argument had some influence but it certainly did not carry the day. That honour went to some observational seismology, presented as a simple picture that most geochemists could understand. I suspect many geophysicists, even those in the mantle convection business, did not really grasp the generality and robustness of my argument.

There was another catch that took some gloss off my satisfaction with the work. The papers were not submitted until 1987. When I moved to the Research School of Earth Sciences (RSES) at ANU I had reluctantly agreed to in initial appointment of five years, renewable until retirement. Effectively I was being asked to give up the tenure I had recently gained in the US and to go back on probation. I was assured that of course if I simply maintained my work then the renewal to retirement (i.e. tenure) would be routine. I was assured also that of course the ANU Institute of Advanced Studies was an elite institution with very high standards, the implication being that tenure at Washington University in St. Louis did not really count, though most Americans would regard it as an excellent institution. I never have been good at bargaining for myself.

The first two papers of the three just discussed were not submitted until April 1987. They were revised in late December and accepted in April 1988. By that time the third paper was submitted. The three were published together in September 1988. So again the review process was protracted. My review for re-appointment was due later in 1987, my five-year initial appointment running until mid-1988. I knew I was cutting it fine but presumed it would not be a major problem. Well, perhaps by then I was aware that it could be a problem.

Ted Ringwood had a history of turning on associates who had come to work with him and got the wrong answer. He was an excellent scientist, in his range of expertise. He had inferred the composition of the mantle that would produce basalts when melted. He was measuring densities and phase relationships at high pressure and, along with S. Akimoto in Tokyo, identified the main changes in crystal structure that account for the jumps in properties through the mantle transition zone.

Ted was also a big thinker and liked to hypothesise grand scenarios on the basis of his lab work. He did his physics and fluid dynamics by intuition. A few of us used to have a game, during Ted's seminars, of picking the moment when he moved seamlessly from things he had measured to stories about how the Earth worked. You had to be paying attention to pick it.

This set the scene for some unedifying academic politics. At the time I arrived at ANU Ted's thing was *megaliths*. He supposed that subducted lithosphere would pile up at the base of the upper mantle, forming what he called a megalith. Various kinds of petrological and geochemical processes would then operate as it warmed up, from which he explained various other observations. Some of the latter parts were not unreasonable, and might well transfer to other dynamical situations. I had by then been actively pursuing the viability of whole-mantle convection for some years. It should not have been a surprise that at some point I allowed that I thought it was not very likely the slabs would pile up to form a megalith at 650 kms depth. Nor did I think the gravity signature he interpreted as evidence would fit. Those were the wrong answers.

I came to understand over some years that Ted had been the one who had pushed most strongly to get someone like me in the Research School. Someone who would complement his theorising by showing how such things would work. I also came to understand that there were other agendas, and there was opposition to my appointment, but that Ted had prevailed, being the dominant (read: bullying) personality in the School.

When I ran afoul of Ted's scheme for the world I became an undesirable. That meant there were two at least who would be just as happy to see me gone. My recollections of the period 1987 to 1990 are a little unclear, but I know that somewhere in there my colleague Ian Campbell and I were talking with a colleague from another School. Ian had said plumes come from the bottom of the mantle (as we'll get too soon) and that was also the wrong answer. Re-appointment committees were required to have someone from another School, and this colleague had been that person, for both of us. He said he recognised that some on the committee were gunning to get rid of us, but he also recognised that the work was substantial. Fortunately there were other allies on the committee.

The committee had several options to deal with my situation. My papers were submitted but not accepted. It ought to have been clear they were substantial pieces of work on a major topic, and continuing what I had been doing when I was hired. I wish I could say the referees would have said they were good work, but the acrimonious mantle convection community could not be relied upon for a generous endorsement (as also witnessed by the protracted review process of those papers). I had published a number of other papers, including with Mike Gurnis, but nothing had come out in the year 1987. Oh dear. So the committee might have decided I wasn't measuring up and shown me the door. Or it might have required that the process be deferred a year so the fate of the papers would be known. Or it might have made a positive decision conditional on the papers being accepted.

In the event, they extended my probation for three years. I was rather shocked and dismayed. I had supposed deferral for one year was a plausible possibility, but thought that approval conditional on the papers being accepted was most reasonable. In earlier times in the School, I learnt from others, the attitude had been to get good people and let them run. Clearly that was not the attitude being applied. I got no kudos for being appropriately ambitious, nor for taking risks. It is pretty clear the hostilities were having an effect. They had not got rid of me, but I was being punished, and discouraged. They counted papers, ticked boxes and looked for possible deficiencies, never mind my overall record. It was not the last time I was to encounter that attitude at ANU.

For some more context, as I had arrived at ANU in 1983 Professor Bill Compston was announcing the discovery of the then-oldest known fragments of the Earth, some grains of zircon from Western Australia measured to be 4.2 billion years old. The detection was made using Bill's ion microprobe, a new and sophisticated instrument that had been quite a few years in development. Getting the ion probe built and working was a risky and expensive undertaking, but Bill had the support of the School and a lot of technical support and money, from the School and national funding bodies. By comparison my work was cheap, and I had no technical support (though it could have helped greatly, then and later). I had made the strategic mistake of taking a risk before I was tenured (at ANU), and the tactical mistake of not going along with the agendas of some small-minded but powerful people. I didn't really think that was why I was there.

I might mention, regarding my mantle convection peers, that the first paper was criticised for being technically questionable (and by implication incorrect) but also for stating the obvious, that the lithosphere dominated mantle convection, so the result was trivial (and by implication correct). I'm pretty sure those attitudes coincided in at least one reviewer's comments. I don't pretend no-one else ever got reviews like those, sadly they are common enough in the academic business.

There were other factors at play within the School that were to become clearer over time. One was the attitude that the real science is measuring, and theoreticians are an optional add-on, if not a waste of money. As I have said, real science is the interplay of both observing and interpreting (with theory, if appropriate), and to champion one or the other is a narrow and limiting view (and often self-interested). Ringwood was one of those who publicly described my work as 'computer games'.

Chapter 12
Some Clarity: Two Convection Modes, Interacting

Sometimes a casual remark can crystallise a thought that has been floating around in your head, partly formed but not fully recognised and spelt out. So it was at a conference once when Brad Hager, in the course of a talk, said 'Plates *organise* the flow'. It clearly came out of his own work on the flow driven by the moving plates, but I saw how it applied to a fully convective flow. Obvious really. Once you have recognised it.

So it was again that my colleague Ross Griffiths dropped a telling remark, probably over lunch. Ross is a proper geophysical fluid dynamicist, whereas I had just picked up whatever fluid dynamics I needed along the way. So Ross casually stated that convection is driven by boundary layers. Well of course, I knew that. But I had not put it at the centre of my thinking. I was close, but Ross' remark got me to the essence. It drew the last shreds of my attention away from other aspects and onto how heat enters and leaves the mantle.

Flood basalt eruptions are the largest volcanic events known in the geological record. In a series of huge lava flows over a million years or so they can accumulate to several kilometres thickness near their centre and spread over an area two thousand kilometres across. Well-known examples are the Deccan Traps in India (Fig. 12.1) and the Siberian Traps in … oh, yeah. Perhaps they are called traps because the lava flows into rifts and other depressions and pools there, but in the large sequences they fill everything up and spread across the landscape. Individual lava flows can be several metres thick and spread for hundreds of kilometres. The Columbia River Basalts are a smaller example in North America, and individual flows can be easily discerned extending for long distances along the sides of the Columbia River gorge. You don't want to be around during a flood basalt eruption.

The Deccan Traps were erupting at the time of the dinosaur extinction 65 million years ago and there has been a vigorous debate over whether the eruptions caused the extinction, or whether that deed was done by a giant meteorite that struck off the Yucatan Peninsula of Mexico during the same time. Current opinion is that the meteorite was the main culprit, but that the Deccan eruptions were already stressing Earth's systems and certainly would not have helped. The Siberian Traps erupted 250

G. F. Davies, *Stories from the Deep Earth*,
https://doi.org/10.1007/978-3-030-91359-5_12

Fig. 12.1 A portion of the Deccan Traps flood basalts in western India. The image shows multiple layers of thick lava flows, ash and other deposits. Image courtesy of Gerta Keller, Department of Geosciences, Princeton University [32].

million years ago, which is the time of the largest mass extinction in the geological record, at the boundary between the Permian and Triassic eras. There is a strong case that the eruption did cause that extinction, possibly because the lavas erupted through carbonate and sulphate formations and liberated a lot of carbon dioxide and sulphur dioxide which caused a global rain of carbonic and sulphuric acids, severely disrupting marine ecosystems.

We met Jason Morgan earlier, as it was he who in 1971 proposed that mantle plumes could explain volcanic hotspot tracks. In 1981 Morgan went further. He noted that in several places a hotspot track seems to emerge from a flood basalt province. For example there is a hotspot track extending south from India along the Chagos-Laccadive Ridge and jumping across the Indian Ocean spreading centre to the currently active volcanic island of Reunion. Morgan proposed that flood basalts are caused by the arrival of a new plume, which starts out with a large *head* that is followed by a *tail* that is the original kind of plume proposed by Morgan. He noted results of laboratory experiments in 1975 by J. A. (Jack) Whitehead and D. S. Luther of the Woods Hole Oceanographic Institute, which showed such head and tail structures.

Whitehead and Luther reported laboratory experiments on the forms of fluid plumes of various viscosities rising from a buoyant layer of fluid into a fluid of different viscosity. If the plume is more viscous than the fluid it rises into then it takes the form of a finger stretching upwards. On the other hand if the plume fluid

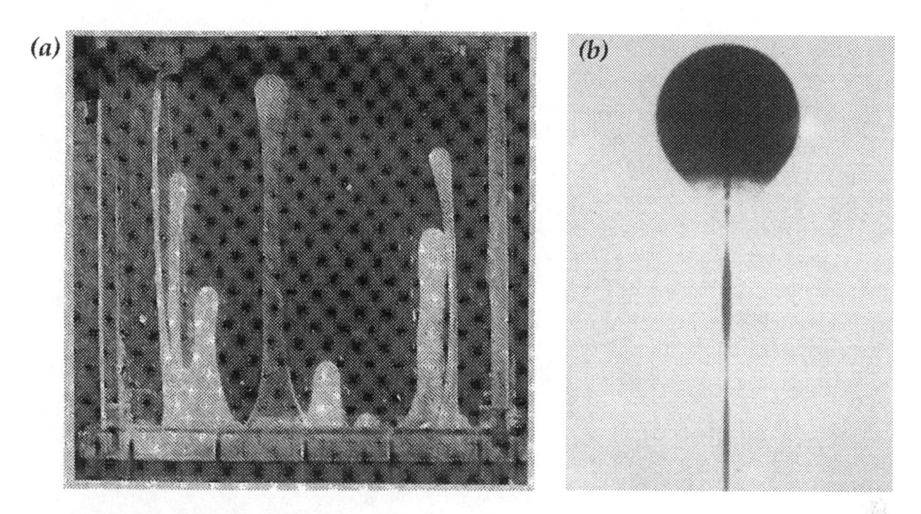

Fig. 12.2 Photographs of laboratory experiments with buoyant upwellings. **a** The buoyant fluid is more viscous than the fluid it is rising through. **b** The buoyant fluid is (much) less viscous than the fluid it is rising through. From Richards et al. [33], reproduced with permission

is less viscous than the fluid it is rising into then it forms a mushroom shape, with a bulbous *head*, followed by a narrower *tail*. These forms are illustrated in Fig. 12.2.

For the mantle we are more interested in the plumes with low viscosity. If a plume rises in the mantle it is probably because it is hot, and its higher temperature will reduce its viscosity. So we're more interested in the head-and-tail structure of Fig. 12.2b. This structure arises because once the head had forged a path through the surrounding fluid it is easy for more low-viscosity fluid to flow up the tail after it, and the tail does not have to be very wide. The case in Fig. 12.2b is fairly extreme, the tails are not usually so thin. As more fluid flows up the tail it expands the head, and increases the contrast between the sizes of the head and the tail.

Others did instructive experiments as well. Peter Olson and H. S. Singer of The Johns Hopkins University in 1985 documented how the head expands and speeds up as it rises. My ANU colleague Ross Griffiths did a series of experiments published in 1986 on isolated blobs, which rise as spheres.

If the plume material is the same as the surrounding fluid, only hotter, a new phenomenon appears: the plume head heats some of the surrounding fluid which then gets entrained into the head. In 1990 Griffiths and our colleague Ian Campbell published experiments in which the plume is fed continuously from below, so it has a tail as well as a head. The image in Fig. 12.3 became quite famous for a time (among those who cared), because of the remarkable spiral that developed in the head.

The spiral forms as follows. As the warm plume fluid rises through cooler fluid of the same kind it warms some of the surrounding fluid close to the plume. That fluid is then more buoyant than the surroundings, and it gets wrapped into the plume head. In the photograph the original plume fluid is dyed, whereas the surrounding

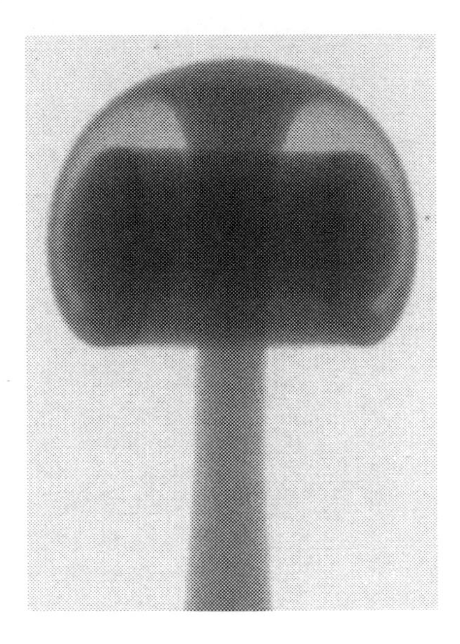

Fig. 12.3 A warm plume fed through an inlet at the base and rising through the same kind of fluid that is cooler. From Griffiths and Campbell [34], reproduced with permission

fluid is clear. The rising head forms what is called a vortex ring, rather like a smoke ring. The warmed surrounding fluid becomes wrapped into the vortex.

When these experiments were scaled to mantle conditions another remarkable result emerged. Because the head is fed by fluid flowing up the tail as well as by some surrounding fluid being wrapped into it, it can become very large. The experiments predicted a plume head might reach 1000 kms across.

A few years later, with a much-improved computational method, I was able to model the growth of a plume from a hot thermal boundary layer at the base of the mantle. Figure 12.4 shows the growth of a plume in more detail. It may reveal the origin of the spiral structure more clearly: the hot fluid rises through the middle of the plume head, spreads across the top and wraps down the sides.

From Whitehead and Luther's experiments it was not clear that a plume head was big enough to account for the larger flood basalts. The work of Olson and Singer and then of Griffiths and Campbell revealed that plume a head could be very large. In the meantime Morgan's idea was revived by Mark Richards, Robert Duncan and Vincent Courtillot in 1989. The proposed sequence is sketched in Fig. 12.5.

Even so there was still debate about whether a plume head could produce enough magma to account for flood basalts. One problem was that a plume head arriving under intact lithosphere would not rise shallower than about 100 kms depth. Melting is inhibited by very high pressures, so the plume has to reach shallow depths to allow *pressure release* melting to occur. Robert White and Dan McKenzie argued in 1989 that flood basalt eruptions were triggered by rifting on the margin of India, which thinned the lithosphere and allowed hot plume material to rise shallow enough to melt in large volumes. However that scenario had problems explaining the rapidity of eruptions, and it would not work for Siberia, where there had been no rifting. Also their version of a plume head was relatively small because they assumed it rose

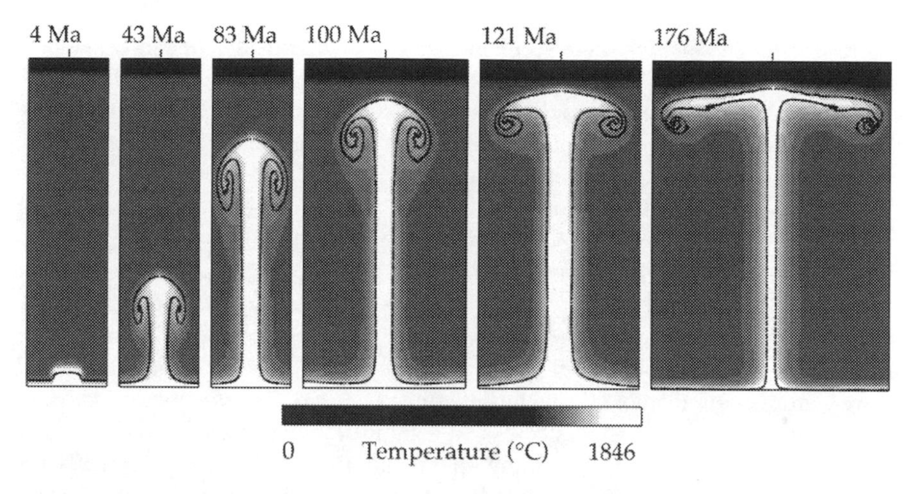

Fig. 12.4 Growth of a plume from a thermal boundary layer, scaled to the mantle situation, except the mantle viscosity does not vary with depth. A line of tracers marks the boundary between fluid from the thermal boundary layer and fluid entrained on the way up. The thickness of the tail varies with time because of small variations in the flow within the thermal boundary layer. From Davies [4], reproduced with permission

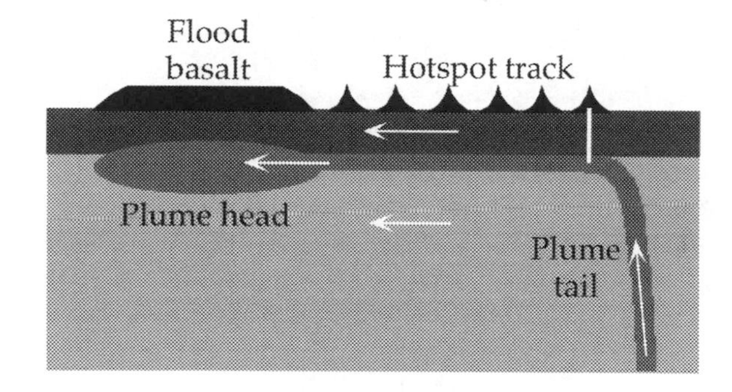

Fig. 12.5 Production of a flood basalt eruption followed by a hotspot track caused by the arrival under a moving plate of a plume head followed by a plume tail. From Davies [4], reproduced with permission

only from 650 kms depth, in other words that convection was confined to the upper mantle.

Ian Campbell had pointed out that plume compositions are probably more prone to melt than normal mantle because, according to Hofmann and White's hypothesis, they will have entrained old oceanic crust, accumulated at the bottom of the mantle, which melts more readily than normal mantle. Some simplified modelling by Matthew Cordery had shown this would increase the amount of magma, but still not by enough.

Eventually more numerical modelling by Alison Leitch and me in 2001, extending models like that in Fig. 12.4, showed that several details combined to increase the amount of melting. First, the right composition had to be included. Second, the plume head needs to be grown from a bottom thermal boundary layer, as in Fig. 12.4, in which case the hottest material flows up the centre of the plume and reaches the shallowest depths. Finally, if the upper mantle is less viscous than the lower mantle by a factor of 30 or so, as was reasonably established by then, it causes the plume head to 'neck down' into a narrower structure that can penetrate to shallower depths more quickly. Alison's models yielded the required millions of cubic kilometres of magma with a million years or so.

Incidentally, this debate provides an interesting example of the interplay between hypothesis, observation and modelling. Morgan's hypothesis that flood basalts are cause by plume heads seemed like a good one, but laboratory models had to include the important combination of features—lower viscosity, higher temperature and the same composition as surrounding fluid—before the resulting plume heads seemed large enough. But then careful consideration of melting seemed to show there would still not be enough magma. Matt Cordery's numerical models included the composition suggested by Ian Campbell, but he started the plume as a uniform-temperature hot blob part-way up the mantle. At the time I concurred with that as a reasonable approach. Only when Alison grew the plume from a thermal boundary layer did the important details of temperature structure become clear, and only when the vertical variation of viscosity in the mantle was included did the plume penetrate to sufficiently shallow depths.

For most of that process it seemed that the hypothesis was not adequate to explain the observed volumes of flood basalts, and therefore that it was not supported by empirical testing. Only when the modelling attained sufficiently realistic detail did it turn out that the hypothesis was supported by the observations.

The lesson is that if a model does not account for observations, you have to look at everything—observations, hypothesis and modelling—to discover whether the hypothesis is really excluded by the observations. The hypothesis was good and the observations were good, it was the models that were meant to express and implement the hypothesis that were inadequate, until they became sufficiently realistic.

The joke about geophysical modellers, among geologists and other rude people, is that if you ask a modeller what is the expected answer from a situation, the modeller will ask 'Well, what answer would you like?' As I remarked earlier, mantle convection modelling acquired a bad reputation early on for models that were too idealised, and they also had too many adjustable parameters. So just to be clear, the details that were included in the above plume-head models were all indicated independently. This was not a case of fiddling with parameters until the answer you want emerges.

Seismology had long revealed that the bottom of the mantle, the lowest 200 kms or so, is anomalous. The nature of this zone, known as the D" layer ('D-double-prime') had also been long debated. The seismic wave velocities increase smoothly with depth through the lower mantle until D", but then they level off or perhaps decrease a little. The seismic waves also undergo some scattering, as though they are passing through some heterogeneities of the same scale as the wavelengths, about 100 km. The scattering suggested that some material of anomalous composition might be present. The decrease in seismic velocity could be due either to a change in composition or to a steepening of the temperature gradient. Some people suggested there might be small-scale convection within D" that would account for the scattering, without appealing to any change in composition. There were also claims to have detected some sharp interfaces in D", and these gained reasonable credibility. A sharp interface would require a change of composition. The seismology had steadily improved over many decades but still it was difficult to get enough resolution to clearly distinguish these possibilities.

Frank Stacey and David Loper in 1983 had considered the thermal structure of deep plumes and of a thermal boundary layer that could be a source of plumes. They had concluded that D" could be a thermal boundary layer that feeds plumes, but the seismological resolution could not clearly establish its temperature gradient and thickness. Still, their use of theoretical approximations was innovative. After Norm Sleep and I had estimated the heat being carried by plumes I realised, in 1990, you could use Stacey and Loper's approximations in reverse, and work backwards from the heat flow to estimate the thermal boundary layer needed to feed the plumes. The result was that a relatively low temperature gradient (1–3 °C/km) sufficed to feed the plumes. This gradient is too small to account for all of the reduction of the seismic velocity gradient. This conclusion supported the idea that there is both a thermal boundary layer and some different composition involved in D". Anyway it gave some insight into the thermal boundary layer inferred to be feeding mantle plumes.

This work also had implications for how plumes would sample the mantle, and we will pick up that aspect in a later chapter.

There was a conference on mantle plumes at Caltech, organised by my old mentor Don Anderson. I think it was in 1991. By this time Don was arguing that plumes don't exist, but all sides of debate were welcomed, and geophysicists and geochemists were there. During the first and third days proceedings were often interrupted by Claude Allègre, who considered that he had the right to speak whenever he felt like it. Sometimes he merely expressed disagreement, or added something he considered important, but at other times he leapt up from the front row and addressed the gathering, taking the floor from whichever unfortunate was trying to enlighten us at the time. The middle day of the conference was more orderly and constructive. The word was that Claude and Don Turcotte had gone to play golf.

Don Turcotte was usually quite civil, unlike his French co-author of the paper on the mesosphere boundary layer I mentioned earlier. I spoke on the third day, presumably explaining how plumes from a bottom thermal boundary layer made geophysical and geochemical sense, though I don't remember exactly. Of course my context would have been whole-mantle convection. The speaking schedule was a bit tight and perhaps I had gone a little over time. The chairperson was clearly anxious for me to finish up. I think there was a question or two, and brief responses from me. Then, from the back of the room Don Turcotte spoke up: 'I've said it before, and I'll say it again. *You're wrong*!' By this time the only response I could make was something like 'I'm not', or I would have been dragged off the stage.

It happened that Richard Kerr, reporter for *Science* magazine, was there covering the conference. A week or two later Richard emailed a draft news article to me and Don. We were portrayed as having had the most unedifying exchange of the conference. Fair enough, at one level. I wrote to Richard explaining that I was usually a bit stressed presenting to an audience that included some unsympathetic members, and I was also about to be given the big hook and didn't have time to respond properly. To Richard's credit he softened the criticism of me, but left his account of Don unchanged. So we had our fifteen minutes of fame, for better or worse.

I presume Don Turcotte was eventually somewhat persuaded by the seismological evidence that was still a few years away, but he moved on to other topics and I don't know what he ended up thinking. Don Anderson's opposition to plumes was a bit puzzling at the time, though it was fair enough to debate them. However his opposition only hardened with time, and I have to say he became more bitter. I found it strange. He did not seem to be the same person who used to provoke us into debates in coffee breaks, often lacing his own comments with his very quick wit.

There was for a time a 'great plume debate', with Don Anderson and Gillian Foulger, of Durham University, as principals of the case against. One of Don's claims was that there was no real physics behind the plume idea, and geologists just invoked a plume whenever they wanted something to heat or rift the crust. There was some truth to the latter part, there was quite a lot of loose thinking about plumes, but the claim that there was no physics eventually provoked me into writing *A case for mantle plumes*, which ended up being published by a friend who was editor of the *Chinese Science Bulletin* in 2005. In that paper I not only reviewed the well-established physics of plumes but argued that plumes are a pretty inevitable feature of a cooling planet with an initially hot core. The great plume debate eventually subsided, but Gillian still runs a web site called, ironically, mantleplumes.org. Sadly, Don is no longer with us.

We have arrived at proposals for two kinds of mantle convection: a plate-related convection like that shown in Figs. 11.2 and 11.3, and plumes such as that in Fig. 12.4. We seem to be getting close to the goal of understanding mantle convection, but there are still some issues to be sorted out. For example, why are there these two kinds of convection, when Rayleigh-Bénard convection (Fig. 9.1) seems to show only one? Why are the two kinds so different? Can they co-exist in the mantle, and if so, how?

Let's look again at thermal boundary layers, mindful of Ross Griffiths' comment that convection is driven by thermal boundary layers. Figure 12.6 compares two kinds of convection. In both the fluid is of constant viscosity, but the heating is stronger than the marginal convection depicted in Fig. 9.1. The strong heating means the thermal boundary layers at the top and bottom surfaces are quite thin, and the rising and sinking columns (sheets, in two dimensions) are also quite thin. The difference between the examples is that left panels are bottom heated and the right panels are internally heated.

There are a couple of things to notice. The bottom-heated sequence on the left has a hot thermal boundary layer with hot rising columns and a cool boundary layer with cool sinking columns, whereas the internally-heated sequence on the right has only a cool thermal boundary layer at the top and cool, sinking columns. The first lesson is that you don't have to have the two thermal boundary layers that are usually shown in Rayleigh-Bénard convection (Fig. 9.1). We've already covered that aspect in considering seafloor topography (see Fig. 11.7). With internal heating you have only the top, cool thermal boundary layer. It would also be possible to make a model with only a hot thermal boundary layer at the bottom—in that case the layer would

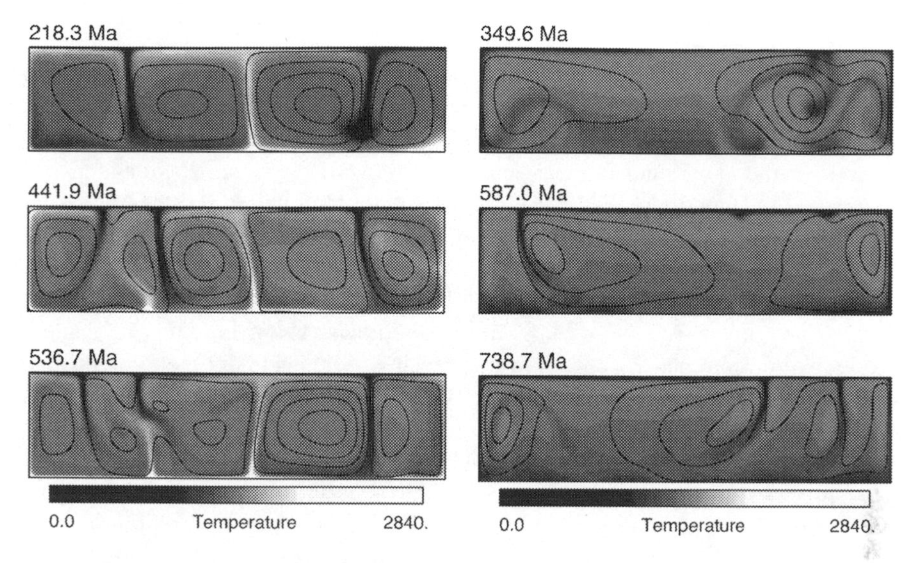

Fig. 12.6 Time sequences of strongly heated convection in constant-viscosity layers. **Left**: bottom heated. **Right**: internally heated. 'Ma' means millions of years. From Davies [4], reproduced with permission

slowly heat up but it would still be thermal convection. So you can have one thermal boundary layer, or the other thermal boundary layer, or both.

The second lesson from Fig. 12.6 comes from the lower left panel. There you can see a cool downwelling and a hot upwelling colliding in the middle of the layer. If you look at the sequence in the left-hand panels you can identify two 'cells' in the right half of the box, but they are a bit unsteady. On the other hand, the left half of the box starts with two cells but they break up in the second panel, and in the third panel it doesn't really make sense to talk about cells. Instead you can just say the flow is rather unsteady, with rising and sinking columns moving about and occasionally colliding.

These models reveal how Rayleigh-Bénard convection is a quite special case, in which there are two thermal boundary layers and the rising and sinking columns are strongly coordinated, so the end result is a nice steady pattern. This case is special because the heating is only gentle—just enough to get the fluid going. The mantle is not gently heated (and cooled), it has a much higher heat flow than that, as do the examples in Fig. 12.6. This suggests that upwellings and downwellings in the mantle need not be strongly coordinated, they might behave rather independently.

So now we can liberate ourselves from the mental straight-jacket of Rayleigh-Bénard convection. There can be hot upwellings from a hot thermal boundary layer at the base, and there can be cold downwellings from a cold thermal boundary layer at the top, and they are best thought of as independent agents that might or might not coordinate.

The mantle, we have inferred, has a cool thermal boundary layer at the top, through which heat flows (into the ocean) at a rate of about 36 terawatts, not counting the continents (tera means there are 12 zeros). There also seems to be a hot thermal boundary layer at the base of the mantle through which heat flows (from the core) at a rate of perhaps 4 terawatts (later arguments might double this). In other words the bottom thermal boundary layer is rather weaker than the top thermal boundary layer. Each thermal boundary layer generates its own rising or sinking columns.

Now a third important lesson. The bottom thermal boundary layer is how heat enters the system, whereas the top thermal boundary layer is how heat leaves the system. In the left-hand panels of Fig. 12.6 the same amount of heat is passing through each boundary layer, but the total heat budget of the system is not the sum of those heat flows, it is the same as those heat flows. The mantle thermal budget (excluding the continents) is 36 terawatts, the amount flowing out the top. The heat input into the mantle is, by our numbers, 4 terawatts entering at the base and, let's say, 20 terawatts of internal heating.

Oops, they don't balance, what's going on? There seems to be a net heat loss of $(36-20-4)=12$ terawatts. It means the mantle is slowly cooling. Very slowly. The rate is about 70 °C per billion years.

OK, we have independent thermal boundary layers transporting different amounts of heat, one in and the other out, but Fig. 12.6 still doesn't look much like plates or plumes. Why are the risings and sinkings in the mantle so different, from each other and from Fig. 12.6? We've already seen the answer. It is because the viscosity, or better the rheology, of the mantle changes with temperature.

The top thermal boundary layer is cool and that makes it stiff, and brittle, so its downwellings are stiff sheets that slowly warm and soften. The bottom thermal boundary layer is hot, and that makes it more fluid, so its upwellings are columnar plumes, with big heads and thin tails. I made the sketch in Fig. 12.7 for my 1990 paper on the thermal boundary layer feeding the plumes, and it still holds up pretty well as a depiction of the main features of the convecting mantle.

Having arrived at this understanding, I determined to write a new account of mantle convection with thermal boundary layers as the central concept, and the different modes and roles of plates and plumes explained. I also determined to discuss every potential obstacle and objection to the resulting picture and to try to prune the dense thickets of confusion around the whole subject. Mark Richards joined me as a co-author in the endeavour. He had completed his PhD thesis with Brad Hager. They had shown that the geoid over subduction zones is an important constraint that requires plates to sink through the transition zone and into a lower mantle more viscous than the upper mantle by a factor between about 10 and 100. This was another strong argument against two-layer convection, but it too failed to be the knockout blow. Anyway our long paper, *Mantle Convection*, was published in the *Journal of Geology* in 1992 and, I understand, became widely used in graduate courses.

The paper begins by showing how a very simple model of a subducting plate could explain the observed velocities of plates. The approach, called a boundary layer theory, had been demonstrated in 1967 by Don Turcotte of Cornell and Ron Oxburgh of Cambridge, UK. This is a bit ironic, because Don became a persistent advocate of upper-mantle convection but we used his theory to show the viability of whole-mantle convection. Their theory was effectively for a more strongly heated

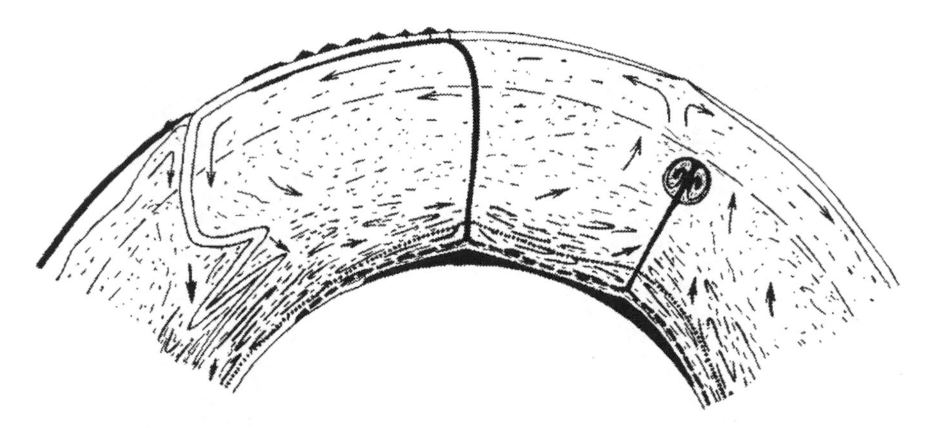

Fig. 12.7 A cross-sectional sketch of the mantle. A plate is forming at the spreading centre on the right and subducting under the continental margin on the left. Continental crust (black) and lithosphere (white) are roughly to scale. The sinking plate buckles as it softens and encounters greater viscosity at depth, eventually spreading and merging in the manner seen in Fig. 11.3. There are denser 'dregs' at the bottom of the mantle, some possibly primordial (black) and others accumulated oceanic crust brought down by sinking plates. A new plume is rising on the right from the thermal boundary layer (dotted), due to heat flowing in from the core. An older plume has generated an oceanic plateau of flood basalts from its head, followed by a volcanic hotspot chain from its tail, as in Fig. 12.5. The plumes draw material from a thin layer (dashed line) of the hottest part of the thermal boundary layer. There are probably also larger 'thermochemical piles' of slightly denser material, not shown in this sketch, see later. From Davies [35], reproduced with permission

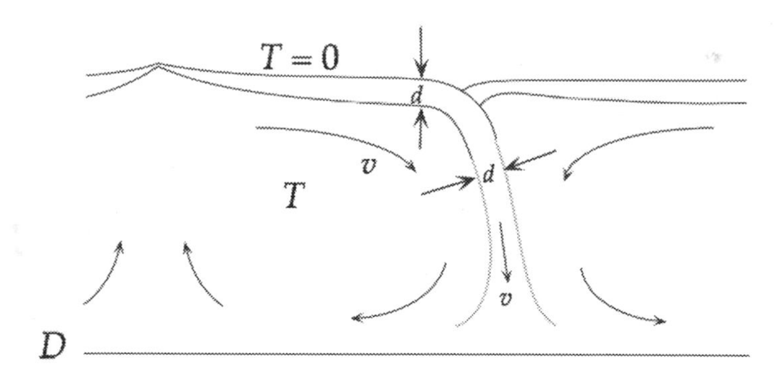

Fig. 12.8 The plate mode of mantle convection, simplified to its essence. Don't worry about the letters, they were used for the algebra to calculate the plate velocity. From Davies [4], reproduced with permission

version of Rayleigh-Bénard convection, with two thermal boundary layers. I invoked only one, and simplified the approximations even further. The simplified picture is sketched in Fig. 12.8.

The basic idea is that there are two opposed forces. The extra weight of the cooler, denser subducting lithosphere drags surrounding mantle down. The mantle, being highly viscous, resists being dragged, and its resistance is proportional to the velocity

Fig. 12.9 Topography of the East Pacific Rise where it is offset by the Eltanin fracture zone. This is the largest such offset in the whole mid-ocean rise system. The tip of South America is visible on the right, and this location can also be seen in Fig. 11.4. Detail of Fig. 11.4

of the sinking plate. There is a velocity at which the two forces balance. With a mantle viscosity of 10^{22} Pa-seconds, appropriate for the lower mantle, the estimated velocity is 8 cm per year, typical of the faster plates.

The plate mode of convection is not really quite this simple, but this little exercise demonstrated that the central idea of a cool thermal boundary layer that sinks because of its slightly higher density seems to capture the essence of the convection associated with the plates. The plates are *part* of the convection, and in fact the major active component. The stiffness of the plates, because they are cooler, traps their negative thermal buoyancy at the surface until such time as the plate reaches a subduction zone, where it is able to sink and drive the interior flow. The plates *organise* the flow because their brittle fracturing determines where the upwellings and downwellings will be, most of the time.

In this way the odd shapes and sizes of the plates are reconciled with the convecting fluid mantle underneath. The plate mode of mantle convection is unusual and distinctive because the active medium, the mantle, changes its rheology from being a deforming, viscous fluid when it is hot to being a brittle solid when it is cool.

Remember how Bruce Heezen thought the Earth must be expanding because there are mid-ocean rises on either side of Africa but there is no subduction in between? Tuzo Wilson and others were puzzled by this too. The resolution of the puzzle is that there are no pervasive, hot, active upwellings under the mid-ocean rises. This is not Rayleigh-Bénard convection. The upwelling is passive. It is just normal mantle flowing in the way sketched in Fig. 12.8, part of the cycle of overturn driven mainly by the sinking slabs.

In fact you can see there is no active upwelling in the topography of the southeast Pacific sea floor. Figure 12.9 shows where the East Pacific Rise is cut by two fracture zones, the larger Eltanin fracture zone offsetting the rise crest by hundreds of kilometres. The sea floor on either side of the fracture zones subsides in the normal manner as it drifts away from the spreading centre. If there were a sheet of hot, upwelling mantle under the rise it would, being fluid, flow past the fracture zones and raise the sea floor on the other side. In other words a fluid upwelling sheet would not be cut cleanly, and it would blur the effect of the fracture zones. It means there is no hot, upwelling sheet, the mantle is normal on each side of each fracture zone. The topography is due to the cooling plates at the surface, not to buoyant fluid mantle underneath.

This means that where ever the plates decide they will pull apart they will pull up whatever mantle is there. If it is normal mantle, as in Fig. 12.9, then they will make normal oceanic crust and normal topography. Of course not everywhere is normal. In some places there is a plume, such as Iceland. In those places the buoyant plume will contribute extra elevation and will make more melt than usual, and of a more 'enriched' composition.

We don't need here to go over all the details of that 1992 paper. It addresses many ideas of how the system, or parts of it, might work. It explains some strong arguments against upper-mantle convection.

Chapter 13
Earth's Lessons: Humility, Power and Science

Studying the Earth can teach us to be humble, as scientists. There is less scope for the arrogance of some physicists that they are discovering the fundamentals, or the truth. Information about the past or the deep Earth is always incomplete, so it is harder to claim you know exactly what is going on. But even Newton and Einstein did not really know what is going on.

Unfortunately the lesson does not seem to carry over into personal humility. There are among scientists a lot of good people who deal calmly and fairly, and also the usual array of personalities: mice, doormats, bullies, bombasts. The trouble is the bullies often wield disproportionate influence. There are also the fearful and the small-minded, some of whom also tend to acquire power, and thanks to them you can endanger your career by being too ambitious.

I will digress for a little while into the politics of doing science, and the process and nature of science, before returning to finish my story of how the mantle might work.

There was a period of several years early in my career when I was interviewing for academic positions. Late in my PhD work I started applying for jobs, and interviewed for a couple, without success. Then I got a postdoctoral position, but a postdoc is a short-term appointment so you keep applying for jobs. After a couple of years I got a faculty position at the University of Rochester, NY, but a year or so into that colleagues started leaving because of the department's financial duress, so I was on the road again looking for another position. The result of all this was that I visited quite a few departments around the US, perhaps a dozen or more, some just to give talks about my work.

About half of the departments I visited had a civil war in progress. Academics tend to be in their heads. They operate by intellectual analysis rather than by emotional connection, empathy or intuition. I know, that's how I lived until my late forties. The trouble is intellectual analysis does not make for good relationships. We humans, it turns out, have a well-tuned set of social responses that tend to keep a small community together and functioning. We would not be here if that had not been true of our ancestors.

G. F. Davies, *Stories from the Deep Earth*,
https://doi.org/10.1007/978-3-030-91359-5_13

The trouble is that intellectualising tends to get in the way of our social responses. We think we are being very rational, unaware of how many of our responses are driven by emotional impulses that we then find rationalisations for. Daniel Kahneman, in *Thinking Fast and Slow* has spelt this out in a way that might get through to more people, by saying we have slow thinking, the 'rational' kind, and fast thinking, manifest as emotional impulses. One of the big emotional drivers is fear. Insecure people need to control the world around them so it is not so threatening. Just look around at political leaders.

In the early 1990s I returned to ANU from five month's leave in the US. It was a much-needed break. My second marriage had broken up and I was walking wounded from that. My standing in the Research School was low: I had been re-appointed after serving my extra three years of probation, but it was clear I was not in favour. I received only a minimal share of the School's considerable resources. The time away was healing. I was treated like a normal human being and a good scientist, because why wouldn't I be? People were friendly. The big department at the University of Michigan in Anne Arbor, where Mike Gurnis was working, and I spent the first three months, was very social.

I returned from the US to find the School in some disarray. The university managers, under pressure from the ideologues to make everything competitive, had creamed money off all the Research School budgets and invited us to compete to get some of it back. The School was bickering over who might write a proposal and on what topic. This was not an open competition for individuals, there would be just one proposal from the School, though I suspect that was not a requirement by the University. The powerful had a scheme but, unusually, other senior people made it clear they were not very impressed. Several other ideas were floating around. I saw how one might bring several of them together under the banner of understanding mantle plumes. A number of people were willing to have their work involved and some colleagues encouraged me.

Over several weeks I put in the leg work talking to people, drafting words, talking again and getting a fairly diverse group of projects fairly well integrated into a story. There were some mutterings from the powers, but enough people were behind it to keep it moving.

At one point one of Ted Ringwood's postdocs appeared in my office and wondered, rather tentatively, if Ted's work could be part of it. I suppose I explained that I didn't really see how it could be related very readily to the other work so it would detract from the coherence of the proposal. He went away. I was also well aware that Ted did not believe in mantle plumes anyway, he thought they were a dangerous heresy bringing disrepute to the School (no, I'm not joking), so why would he want to be involved?

A week before the deadline for submitting proposals a meeting was convened of all those involved in the proposal—except me, the main author. It was proposed to dump my proposal and develop another. My colleagues said that wasn't feasible, there wasn't time and there was no other coherent alternative. It was a very obvious move to cut me out, whatever else might have been going on, but it didn't work. So the School had no choice but to bless the proposal and submit it.

The proposal was funded. It brought half a million dollars into the School in perpetuity, added to the permanent funding base. Hallelujah, I had done something meritorious in the School. A number of colleagues would directly benefit and the whole School would be better off. It had happened because of the insights I had developed in my own work, which enabled me to integrate my colleagues' expertise into a good story. It was a turning point. Finally I felt like a full member of the School.

Well, no.

Ted Ringwood was desperate for some funding to continue two of his postdocs. He demanded something be done. A well-connected but unreliable colleague, who always bent with the wind, said they were planning to re-assign the money. I said the university rules didn't allow that and thought it was unlikely.

I was included in the meeting at which the issue was to be raised, because I was a representative on another University body, but the meetings clashed and I only got to the School meeting part way through.

I walked into an ambush. Ted was given the floor. He ranted and accused. He said I had deliberately excluded him from the proposal, out of personal spite. It was standard form for Ted to make personal attacks. I was never any good at this kind of conflict, and the group was routinely unreceptive to my points of view. In this case I was stunned as well by the gall of the claims, their untruth and the fact that the School was planning to flout the university rules applying to the grants. One might have called Ted on his presumptions and wild accusations, which were quite unfounded. One might have told Ted if he really wanted to be part of the proposal he could have had the guts to come and talk to me himself, instead of sending a junior flunkey who was incapable of getting to the nub. Of course I did none of that.

Colleagues who were part of the proposal and/or who I thought were allies kept their heads down, or even argued that it was very reasonable to apply the funds to Ted's work, which was rubbish. This was standard form with most of my colleagues. Keep their heads down and never mind what's going on.

A big chunk of the money went to Ted. My colleague Ian Campbell, whose work was central, got nothing because he was the chief heretic. I was bought off with funding for a junior research assistant, whereas I could have used someone more senior with serious expertise in computation. Other colleagues got their bits.

I was completely disgusted with most of my colleagues. I basically told them to get stuffed and retreated to my corner. My triumph had turned to ashes.

Ian Campbell and I went to talk to a University manager about the rules being broken. He listened somewhat sympathetically to our plight, even allowed it was not how he would run a School, but senior management declined to get involved in School affairs. What then was the point of all our effort? What was the point of the university soliciting proposals and judging them if Schools could just spend the money however they liked?

Ted heard we had gone to higher levels and accused us of *Disloyalty*. We should be reprimanded or severely punished. (Who was it broke the rules?) I was told he was even threatening legal action until talked out of it.

My self-imposed isolation within the School lasted a good six months. Eventually the colleague who had I thought was my most enduring ally, but who had welched in the meeting, came and requested that we move on, it was not good to have such hostility within the School. Well yes, but he didn't say that to Ted and other powers did he? I relented a little and things slowly eased, but of course nothing was resolved and my standing in the School was not restored.

I do wonder, looking back, why I persisted. I know the reasons, I explained earlier why I had returned to Australia, but it was toxic. I give myself some credit for surviving it without developing a serious illness. You can't live with that level of anger and frustration for long. The stupid thing is it wasn't finished, there was more to come, though not so much from within the School. That was the lowest point though. Around that time my personal life took a new turn that increased my self assurance and self awareness, so I would be better able to handle the nonsense, which was just as well.

So that was one of my experiences of academic civil war. At least one of the departments I had visited early in my career had split, which is a shame. I have had palaeontology colleagues, for example, whose work had little direct relationship to mine, but which I found fascinating anyway. That experience was beneficial and fed other interests of mine. But conflict is widespread in academia. Of course it is widespread elsewhere too, stories of nasty office politics are pervasive. We are not, in our modern Western culture, very good at cultivating emotional maturity. If your family did that for you, you are fortunate, but many of us did not get it.

As I have said, studying the Earth can give you some good insight into how science works. It is not an inexorable garnering of facts and laws. It is not about truth. It is about finding descriptions of the world that provide useful guidance to how the world works. I have for some time involved myself in trying to understand economics, in general terms, and economies, a very different topic. The Earth is messy but economies and human societies are messier. Through this I have sought to figure out how economics might be done scientifically—not, I hasten to add, by treating an economy as a clockwork, that's the trouble with existing economics. Rather, by proceeding systematically to try to learn whatever important features might be gleaned by careful observation. It requires a more general conception of science than one might reach just from doing laboratory experiments.

Here is my best general statement of the *process* of science.

> I think the essence of science is the *perception of patterns* in our observations of the world, the careful *description* of such a pattern, the *deduction* of *implications* from that pattern, and the *comparison* of those implications with further observations.

You might notice I did not use words like *fact, law or prediction.*

Let us take a familiar and perhaps seemingly trivial example to illustrate this process, with apologies to Ptolemy. We may notice that about every 24 h the sun rises in the east, moves high across the sky and sets in the west. In other words, we notice a pattern in the way the sun appears to move. We might then formulate a hypothesis to describe or encapsulate the pattern. For example, we might hypothesise that the sun moves steadily *around* the earth once every 24 h on a circular path that carries it sometimes above the horizon and sometimes below it. From this hypothesis we can deduce that the sun should rise again in the east, about 24 h after the last time we saw it rise. We can wait and see if this happens. If it does, we can consider that our hypothesis is supported by a new observation, and that it seems to be a good hypothesis. If the sun continues to rise about once every 24 h, we might conclude that our hypothesis is supported by observations of the real world, and start to call it a theory.

Although it may seem trivial, this example already allows several important things about the scientific process to be drawn out. The first stage of the process is the perception of a pattern in the world and the description or formulation of this as a hypothesis. This process of perception and formulation is sometimes called *induction*. However despite this formal-sounding name, the *perception* process *is not a rational process*. It is a process of cognition that is deeply wired into our brains and has nothing to do with logic. Our clever brains are very good at perceiving regularities or patterns in the world, such as similarities in the shapes of animals, trees or faces, or regular events like days and seasons or musical rhythms.

Indeed, we often perceive different patterns in the same observations. For example we can see faces in clouds, and there are clever visual puzzles that have been constructed to look first like one thing (a young woman's face) then like another (an old woman's face). In such cases our brains are receiving the same signals from the world, but we can make different stories from them, depending on how we 'look' at them.

The intrinsic ambiguity of perception means that we have to be very careful about claiming the pattern we perceive to be the 'true reality'. This is not an obscure point of philosophical debate or of optical illusions, it is of central importance in science.

For example, Einstein's theory of gravity is not a modification or extension of Newton's theory, it is based on quite different conceptions. Einstein abandoned Newton's idea of force acting at a distance and replaced it with the idea of local variations in the rules of geometry. These are entirely different conceptions of how the universe works, though they yield similar predictions in many circumstances. So how can we say that Newton's conception was true, or false, and can we have any confidence that Einstein's conception will not be replaced? Where, then, is the 'truth'?

The second stage of the scientific process is the *deduction* of consequences of the hypothesis, and the comparison of those deduced consequences with more observations. This deduction and comparison stage is often called the *empirical testing* or just the *testing* of the hypothesis. In contrast to the induction stage, the deduction stage is logical and rational. Deducing the consequences of our hypothesis about the sun's motion is rather trivial and doesn't require any sophisticated logical tools, but it is nevertheless a strictly logical process.

It is from the deductive stage that science has gained the reputation for being rational. It is also from this stage that much of its reputation for being impenetrable derives, since very elaborate logical methods are often used. Deducing the time of next appearance of our orbiting sun is simple, but deducing the dynamical consequences of a phase transformation deep in the mantle is not simple: it requires the help of sophisticated mathematical tools and possibly of computers. Mathematics comprises a vast collection of elaborate logical structures already worked out, which is why it is a very useful and frequently used tool in this stage of the scientific process. Computers extend our ability to use mathematics, by allowing us to 'solve' equations that otherwise are intractable.

Now, we have looked briefly at the *process*, but how do we judge the *success* of a theory? The criterion for judging a hypothesis or theory is that it is *useful*. Our simple theory of the sun circling us is useful, because we can predict more-or-less when and where the sun will be. However there is another theory, that the sun is stationary and the Earth is rotating. That's a bizarre idea. It means if I stand facing east the ground ahead is continually tipping away from me, tumbling headlong.

How do we tell which theory is true? That's not a very helpful question. The better question is which theory is more *useful*, because the point of a theory is to give us useful guidance about how the world works. The answer at this stage is that either theory works as well as the other.

Observations over a longer period will reveal that the sun's circles around us are not fixed, but move slowly north and then back south over the course of a year. We could add that to our description of the sun's pattern of motion. Then there are the stars and planets. The planets (and the sun) move against the background of the stars, so we can give them their own circles. But then their motion is not uniform, sometimes they get ahead, sometimes behind, and sometimes even reverse their motion for a while. So we can add circles upon circles, epicycles. Ptolemy's description of the heavenly motions was a culmination of this approach. It works quite well. It is quite a useful description of the motions of the sun and planets, though it is rather complicated.

Ptolemy's theory I would call science, though it is a rather unwieldy theory. Imagining different forms of mantle convection and running them in a computer is not necessarily science, unless it is part of a process of learning how things might work so they can be compared with what we can observe, without being able to see very well into the mantle. Postulating that people are rational calculators, have no social interactions and can see into the future, and building a theory of an economy on that basis might also be part of science if it is part of a process of discovering what are the important factors in a real economy. However if such a theory is simply held to be a beautiful or gratifying theory and it is used to advise governments then I would say it is not science, it is part of a belief system. Unfortunately that is currently how much of the world is governed.

Copernicus proposed an alternative to Ptolemy's scheme. He proposed that not only is the Earth rotating on an axis but that it moves around the sun, and that the other planets also move around the sun (but the moon does not). At the time Copernicus' description was simpler but hardly any more accurate than Ptolemy's, and observations did not give a strong reason to prefer one over the other. That is according to Thomas Kuhn, whose first book in 1957, *The Copernican Revolution*, is well worth reading for an understanding of the complex interplay of observations, hypotheses, culture, myth and religion that is often involved in science.

The more accurate and systematic observations of Tycho Brahe were required to reveal more clearly the limitations of Ptolemy's theory, but there were discrepancies with Copernicus' as well. Kepler, after arduous search, replaced Copernicus' circles with ellipses, and was then able to arrive at his three 'laws' of planetary motion. These further reduced the discrepancies, but did not eliminate them.

In turn Newton eventually replaced Kepler's rules with a simpler one again, that all bodies are attracted to each other by a gravitational 'force' that varies inversely as the square of the distance from the body. That is another bizarre idea, bizarre enough to trouble Newton himself, even though he was also heavily into alchemy and astrology. (Could he have conceived of gravitational 'action at a distance' without astrology's idea of planetary influences?) The remaining discrepancies with observations were accounted for by the planets pulling on each other. Newton's theory triumphed because it could explain the motions of planets, falling apples and projected cannon balls.

Except, there was still one small discrepancy. Mercury's motion does not quite fit Newton's predictions. Einstein, coming from quite different concerns about the speed of light being the same in all directions, regardless of the observer's motion, proposed that the rules of geometry are modified in the vicinity of massive objects. Einstein's theory can explain Mercury's motion. It also predicted that the path of light would be bent near the sun, and this was later observed. Einstein's theory has triumphed, and it explains the expansion of the universe, black holes and many other exotic phenomena.

Except ... Einstein's theory is quite incompatible with quantum mechanics, which grew up in parallel and triumphantly explains the behaviour of very small things.

This bit of history reminds us of some important things about science. Scientific theories are contingent—they are always liable to being replaced by a 'better' theory. Science proceeds by the interplay of observation and theorising. Theorising is a *creative* process, not a rational deductive process. Science does not proceed inexorably. It moves by fits and starts, depending on key things coming together in one person's mind. (Wegener became intrigued by similarities of continents with large ocean spaces between them. Wilson puzzled over why some kinds of faults seemed to just end, and it led him to think more about how brittle lithosphere might behave.) Observations always involve some inaccuracy, some 'uncertainty'. Sometimes science is stimulated because of a new or more accurate measurement (observation). Observations might also be rather incomplete, as our picture of the mantle deduced from seismic waves was at first quite fuzzy, with large un-probed regions, and has gradually been sharpened and filled in.

Which of the above theories is 'true'? It can't be Newton's, because Newton's theory gives the 'wrong' answer for Mercury. It must be Einstein's, but Einstein's theory is 'wrong' for atoms. And Newton's theory is still very good for describing falling apples, cannon balls and most of the planets. Is 'reality' a fixed Earth, or a fixed sun, or forces acting at a distance, or twisted geometry, or the weird probability waves of quantum mechanics?

These are not useful questions, because they misunderstand the purpose of a theory. The purpose of a theory is to give us *a useful guide to the behaviour of the world we observe.*

By that criterion, all of these theories are useful, within a certain context, and to a certain accuracy. Ptolemy's theory explains the daily and seasonal motions of the sun, to a reasonable approximation. The Copernicus-Kepler-Newton account gives a more accurate description of the apparent motion of the sun, and covers the planets to quite high accuracy too. (Notice that I interchanged the words 'explain' and 'describe' in the preceding sentences. Scientific 'explanation' is really just 'description', though often of a detailed and precise kind.)

The test of a theory is whether it gives a *useful description* of *what we can observe.* Whether a description is useful depends on the context, including the accuracy we regard as useful in that context.

We can leave the question of what 'reality' is 'behind' our observations to metaphysicians and theologians. Unfortunately many scientists became enamoured of the idea that science is in the business of 'reading the mind of God'. It's another distraction, unless God's mind is very changeable and context-dependent. So too can we leave the question of 'truth' to others. It is apparently a shocking claim to many people that science is not in the business of revealing Truth. Rather, science is, to emphasise the difference, in the business of inventing *useful stories*, stories that may be rather loose or may be very precise.

The theory of plate tectonics became widely accepted because several kinds of observation were found to be highly consistent with it and not consistent with alternatives: earthquakes on mid-ocean rises and elsewhere, magnetic stripes on the sea floor and sediment ages versus distance from rise crests, as we saw in Chap. 8. Many other observations are also consistent with it, though often less decisively distinguishing it from other theories.

A detailed theory of the physics of plumes in the mantle is consistent with key observations of hotspot volcanism, hotspot swells around that volcanism, and many details of the chemistry of the erupted rocks. Plume theory is scientific in the sense of providing a useful description, though observations distinguish it less decisively from some other ideas, and it is not quite as widely accepted as plate tectonic theory.

The theory of a plate-scale of mantle convection driven by the weight of sinking sheets of lithosphere and extending through most of the depth of the mantle is also consistent with important physical observations and is fairly widely accepted, at least in broad terms. Whether the chemistry of rocks from the mantle is consistent with it is still a matter of lively debate.

If you want to prosper in academic research then you should become known as an expert in a particular topic. You don't have to make earth-shattering contributions, but they should be widely recognised as solid work. You should have a small empire of students, assistants and postdocs. It is then good to rise up the management hierarchy, at least some of the way. You should also get yourself on committees, both within your institution and within your profession. The latter activities are likely to attract awards and prizes for your contributions to your field. Recognition will start to compound and you will become prominent in your field, though you might not have done much actual research for some time.

I do not mean that summary to be totally cynical. I have known good people who have followed that path. I have known people of modest academic accomplishment or limited people skills, or both, who have also followed that path. I have known people who have gained recognition for academic accomplishment without rising into management. However I would say that management is the surest ingredient for recognition and certainly for a pay rise. I think there is a simple factor involved in this reality: the rules are written by managers, so the rules reflect the perceptions of managers.

In this environment it may not pay to be too adventurous. The rules favour managing over lone-wolf creativity. Competitive funding exacerbates the tendency. Whatever the guidelines might say about grounds for promotion, for example, many managers will want to minimise their own risks by ticking boxes rather than approving some outlier. Pursuing the harder problems can reduce your publication rate. In an acrimonious field you may not get unqualified endorsements. Einstein was not a dean and he did not run a little empire, he worked in a patent office until he had one great accomplishment that was quickly recognised. He was fortunate his work was recognised, but he still worked more as a lone wolf, and he was fortunate to be allowed to. In these more timid and managed times, you really can endanger your career.

My problem was I didn't stay in my defined field. I strayed outside my job description. I kept pursuing big questions. It was bad enough that I went trampling around in the field of geochemistry where I didn't belong. It was worse when I got interested in economics.

In physics, if you pursue a big question, like fundamental particles, you become hyper-specialised. In Earth science, if you pursue big questions, you must become a generalist. Plate tectonics impinges on many other aspects of geology. The theory of driving plates has to be consistent with what is known in many sub-disciplines, and some of your key evidence might come from those sub-disciplines. Geochemistry was the outstanding example.

A difficulty in trying to be a generalist is that it is not hard to get some details wrong, and the specialists will be all over you if you do. On the other hand the specialists will not really understand the parts of your arguments that come from other specialties, so their overall impression will be incomplete and may be negative. Even if they don't have specific criticisms, they may simply have little basis in their own experience for judging your work and may give it little regard as a result. But I persisted in looking into the geochemistry, as I will recount in the next chapter.

In the meantime I had a long-standing interest in why the world seems to be so badly governed. I had read a book that explained the theory of free markets. It became abundantly clear that the theory of free markets is so abstracted and idealised that it has very little relevance to real economies, yet its concepts dominate the world. Later I read the book *Complexity* by Mitchell Waldrop. It includes stories of economists exploring new approaches. I recognised that this was the way to make sense of an economy, and it implied behaviour that was radically different from that implied by free market theory. I was quite excited by this revelation. There were economists who understood this, but they were marginalised in the profession. It dawned on me that as a scientist I could explain why the mainstream theory was nonsense, in very specific terms rather than as a general complaint about the world, and the profession had no hold on me. Nor, of course, need they pay me any attention. On the other hand it was an intimidating prospect, because I would have to take on, debate, people secure in their roles and standing, people commonly much better at verbal jousting than me, and so on.

I looked for a way to do this, and in 1999, having completed my book *Dynamic Earth*, I decided just to give it a lot of my time. It became a very big job, learning a great deal about a very different field. By the end of the year I had a large manuscript. I had decided I needed to include all the arguments needed to bolster a very different approach to the subject—an incomplete case might just be brushed aside.

There was a bit of risk with neglecting my science for a while, but that had been coming along well and would again. I reasoned I had one of the more privileged jobs in the world and the best way I could repay that privilege was to critically analyse a very important subject from an important perspective (that of a proper, practising scientist), never mind what my official job description was. Universities are (or were) supposed to be about coming up with new knowledge and sharing it, and that was my plan, so my own conscience was clear. Of course these days universities are much more degree factories to feed the needs of The Economy, but that's not how it ought to be and the old ideal wasn't completely dead. I was at this time still *persona non grata* at work, though we finally had a new Director who was not overtly hostile, though he was not exactly supportive. I figured he didn't want to know about me and I didn't want to know about him, so I'd work away quietly in my corner. I didn't advertise what I was doing to my colleagues, but I didn't deny it either. A few eyebrows were raised but nothing much was said.

That manuscript eventually became *Economia*, published in 2004. It has, as I say, a small and dedicated following. I have written quite a lot more since, especially after my retirement in 2010, including some quite short versions. It is possible, I concluded, for our societies to function in a way that is much less destructive of the natural world and much more supportive of people.

I had, in 1996, applied for a promotion at work. Colleagues of comparable vintage elsewhere were moving up the ladder and I thought I would test the waters. I was not successful. In a de-brief interview I was told I was doing good work but I needed something to lift me above the crowd.

In 2001 I tried again. By then the book *Dynamic Earth* was out and had received good reviews. I was continuing to extend my research, this time by including mantle chemistry in my mantle convection models, but it would not be published until the next year. Also, and somewhat to my surprise, the Director of the time had become rather more positive towards me and my work, and he encouraged me to apply. His word might add some weight to my case.

Again my application was not successful. This time, however, it was made known to me, indirectly as I recall, that I would have succeeded except the University had a quota on the number of promotions in a given year. They had had to choose between me and one other candidate, and they had gone with the other one. This was disappointing but also encouraging. Evidently I was pulling clear of the crowd. It is conceivable that my lesser attention to my science in 1999, manifest as no papers published in 2000, though there had been four each in preceding years, might have tipped that balance, but nothing was said to that effect and I have no reason to think they were taking such a narrow view. The problem was the quota.

By the way, the quota on promotion to Professor was regarded by the powers of the University as a significant concession. Before that there was no 'internal promotion'. If a Professorial job opened up you were free to apply for it. The conceit was that they would search the world and appoint the best person in the world. It's always more complicated than that of course, not least because Harvard, Cambridge and their ilk are out there claiming to do the same thing. In the event, it was conceded the next year that in an age of non-discrimination they couldn't really justify a quota and it was abolished. This would have preempted the possibility of them being sued.

By 2002 the Research School had a new Director. He was a new experience for me. He just assumed I was there because I did good work. He was not part of the old-boys club of the School, and said so in so many words. He told the School he thought we were demoralised (he was right there) and he would support anyone who just got on with good science. (The internecine wars had continued in various forms, though I was less involved. Political, ideological and financial attacks on the University had continued.) The Director made clear to me that I was included. Gosh, a boss who supported me, for the first time in nineteen years.

The funding rules had been changed, in accordance with the ruling ideology. Money had been removed from the University and we were to compete within the Australian Research Council for some of our research funding. The Director was used to this kind of regime in the US and he got the whole School moving to get proposals submitted. It paid off, our School topped the funding awarded across the University. Even I got some funding (there were few people doing my kind of work in Australia and it was not highly regarded in the broader geology community). For the first time I had some independence in running my work, beyond an office and a modest computer. I hired Research Fellow Jinshui Huang, whose good work I will describe later. We clarified some more basic controls on the geochemical results and observations, and confirmed that the general results still applied in three-dimensional flows.

The signs were all good for another go at promotion in 2002. My book was out, the new work was coming out, the quota had been abolished and the Director was unequivocally on my side. He encouraged me to re-apply, which I did.

My application was not successful. This promotion committee did not consider that I was of sufficient merit, never mind what the previous committee might have said. I was floored, again.

After all the recent positive indicators, after all the petty and vindictive treatment over many years, after all the work to get some traction in a field (mantle convection) that could be rather petty and vindictive, it was a heavy blow. It seemed I just could not shake the monkey off my back.

It was possible to appeal, but only on grounds that proper procedures were not followed. Since most procedures happened behind closed doors that was virtually impossible to establish. I appealed on the grounds that I had been considered of sufficient merit in the previous round, that the quota in the previous round had been unfair, and that the University was unreasonably inconsistent in its decisions. The appeal was rejected of course. No explanation, no recourse. Except perhaps legal action, but I knew that would be an immense and draining undertaking and they of course would hire crack lawyers to defend themselves. So I dropped it and returned to my state of resentment with the University, feeling that my situation was still affected by all the previous conflict within the School.

Chapter 14
Some Chemical Clarifying

There was a remarkable change of opinion among mantle geochemists in 1997. I attended two conferences in the US six months apart: in December 1996 and June 1997. At the first all the usual attitudes, debates and conflicts were evident regarding whether the mantle is layered or not. I had been arguing for nearly a decade that heat flow and topography were strongly against layering, but that argument had little effect. It required four sentences and some high-school physics to appreciate it.

Early in 1997 a paper appeared by three seismologists, Steve Grand of Texas and two of my ANU colleagues, Rob van der Hilst and his student S. Widiyantoro. They had imaged the mantle in three dimensions, and had detected subducted lithosphere extending down from several subduction zones. They coloured it red and showed a picture in which the big red slash clearly extended deep into the lower mantle. There in a single image was the proof of flow passing between the upper mantle and the lower mantle. That picture was certainly worth a thousand words.

The conference in June, 1997, was the regular big international geochemistry conference, the Goldschmidt Conference, in Tucson Arizona. To my amazement most of the mantle geochemists were shrugging and saying there must be whole-mantle convection. They had seen the seismologists' picture. They didn't like it, they didn't know quite what to make of it, but they conceded.

Well, kind of. Perhaps there was a signal from the core. Perhaps the layering was intermittent. Perhaps only the slab went through, and there was still not a lot of mixing between the putative layers. Anyway whole-mantle convection was now the default and the layered mantle case required special pleading. Of course few geochemists showed any awareness that strong arguments had been around for a while.

A few geochemists simply ignored the evidence and carried on modelling a separate upper mantle, Gerry Wasserburg and some of his students prominent among them. I found that my ideas and models still were not readily accepted. It seemed they just could not understand how long-lived isotopic heterogeneity could persist in a convecting mantle. Their intuition was that mantle convection should homogenise the isotopes, like stirring cream into coffee. No amount of showing pictures and

G. F. Davies, *Stories from the Deep Earth*,
https://doi.org/10.1007/978-3-030-91359-5_14

explaining seemed to get through. They still wanted a hidden layer in which to store stuff for a long time.

Uli Christensen was a creative modeller of mantle convection, generally seeking to go beyond the more simplistic approaches that gained the subject a bad name. Geochemist Al Hofmann was more open-minded than most in his approach to interpreting the data from the mantle. With Bill White, he had proposed in 1982 that subducted oceanic crust sinks to the bottom of the mantle and is eventually sucked up in plumes to impart distinctive chemical signatures to ocean island basalts erupted at volcanic hotspots (Fig. 10.5). In 1994 Uli and Al got together and modelled the Hofmann-White hypothesis. This was an important step in exploring the relationship between mantle convection and mantle chemistry.

Previous studies had looked at the stirring of tracers by convection. The distinctive features of the new study were that the formation of oceanic crust was explicitly simulated, the resulting differentiated sandwich was then fed back into the mantle, the tracers of subducted oceanic crust were negatively buoyant and that the isotopic evolution of the crustal material was simulated. They found that around 15% of the tracers accumulated at the bottom of the model in 'pools' or piles where the horizontal flow converges and turns upwards. This pooling increased the residence time of the tracers to about 1.4 billion years.

The neodymium and lead isotopic evolution of the tracers was modelled and plotted in standard geochemical formats. They found that plausible parameters could reasonably reproduce the observations, like those in Fig. 10.4. Although the tracer ages averaged only 1.4 billion years, the slope of the computed points in the lead diagram implied an apparent age of around 2 billion years, because the plot tends to be controlled more by the oldest data.

It was a striking accomplishment to carry the modelling through from convection to isotopic plots, a laudable example of doing real science instead of just exploring models. The study boosted the plausibility that whole-mantle convection could account for the ages and diversity of mantle isotopes. It demonstrated that pooling at the bottom did indeed increase both the residence times and the diversity of isotopic compositions.

Nevertheless there were of course aspects of the models that could be improved. Their models were not as strongly heated as the mantle, and as time went by computers were capable of more resolution. Their heating was 80% from below, which caused strong upwellings and an unsteadiness in the flow that might exaggerate the efficiency of stirring, as our earlier modelling had shown. The flow had a subduction zone that migrated rather faster than real subduction zones.

The models might not have been run for long enough. They were run for an equivalent of 3.6 billion years at present flow rates, but the mantle would have been running faster back then.

Christensen and Hofmann acknowledged that the way the tracers were sampled affected the spread of points in the synthetic isotope plots: larger sampling volumes reduced the spread. The same would be true in the mantle: the more melting and the more mixing of melts the less variation there would be in the observations. As well, not all of the geochemical parameters were well-determined independently.

This meant that the spread of computed isotopic points was not a robust feature of the models and the spread of the observations depended on melting and segregation processes that were not well constrained. The more robust features of the observations are the ages, or apparent ages, and the differences between MORB and OIB observations.

The layered mantle got a new run in 1999 when a pair of papers appeared in Science magazine proposing a separate layer deeper in the mantle, about the bottom 1000 kms of its 3000 km depth. The first paper, by Rob van der Hilst and K. Kárason, offered seismological evidence that might indicate the presence of such a layer, though it was circumstantial rather than compelling. The second paper, by Louise Kellogg, Brad Hager and van der Hilst, proposed a layer only slightly denser than the overlying mantle whose upper interface would, as a result, undulate in space and time. This might not be inconsistent with deeply plunging slabs of lithosphere, and the non-horizontal interface might not give strong seismological signatures. Although some of my former colleagues were involved in this, colleagues who had been prominent in arguing against the old layered mantle, I found the case quite marginal from the beginning.

The papers caused a lot of excitement. Geochemists could have their hidden reservoir. Seismologists could set about finding the subtle signals from the putative layer. Modellers flocked, because many of them knew how to include a layer with a wavy interface. Many models were published showing how such a layer might work. Geochemists relaxed into their old habits. But the seismological evidence for the existence of the layer never seemed to get any better. The idea gradually faded from prominence, though it remained as a resort for geochemists. There was one clear argument against it.

I argued, whenever I had a chance, that my original argument against layering, the topography that would be generated, applied to this layer with nearly equal force, because it would have to contain around half of Earth's heat sources, so plumes and plume topography would be much more prominent. The argument never seemed to carry any force, even among geophysical colleagues. I had to carefully qualify my later work, for example by saying the *only clearly resolved* features in the mantle were the D" layer in the bottom 100–200 kms and the 'super-plumes', better called super-piles, under Africa and the Pacific. We'll get to those shortly.

By 2001 we could improve on the models of Christensen and Hofmann. I had in the early 1990s converted to a computational method called multi-grid that eventually I got working much more efficiently and robustly than my earlier methods, though it was still a bit cantankerous and often had to be coaxed to do the harder cases. I had used it in 1995 to explore the interaction of subducted lithosphere and plumes with phase transformations in the transition zone, showing that both were reasonably likely to penetrate the zone, thus supporting whole-mantle convection.

I determined to have another go at the geochemical modelling. With a better computer method and more powerful computers I could do the models with full mantle heating. I put simulated plates on the surface with slowly migrating subduction zones, so the flow would be of the right kind, with the right kind of time variations. I made the radioactive heating internal to the mantle, rather than entering from the

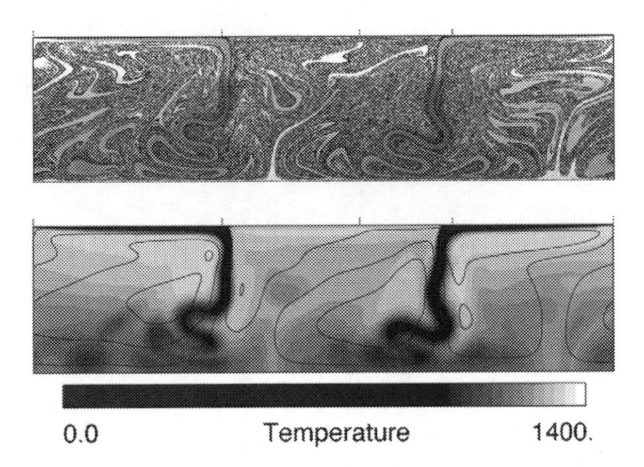

Fig. 14.1 A well-processed model mantle after 4.2 billion years (Earth time). The bottom panel shows flow lines and temperature (as grey-scale). The top panel shows the distribution of tracers, representing old subducted oceanic crust. In this case the tracers are passive, of neutral buoyancy. Two subduction zones can be seen. As fluid rises to the surface near the ends of the box tracers are removed to a very thin layer, simulating oceanic crust, leaving a layer devoid of tracers, simulating depleted mantle. That 'sandwich' is subducted and stirred by the flow. The folding of subducted lithosphere at depth is similar to the sketch in Fig. 12.7. From Davies [36], reproduced with permission

core as Christensen and Hofmann had. I sampled the mantle more in the manner of plates and plumes: plates continuously sample the top of the mantle and plumes sample the bottom. I avoided calculating geochemical-style diagrams, focussing on mean ages of simulated MORB and OIB samples.

Most importantly, I ran the models for much longer. We had realised in Mike Gurnis' thesis work that the mantle would have convected a lot faster in its early days, when it was hotter, than today, so it would have 'turned over' many more times than if it had just run at present rates. Mike had come up with a rough conversion from 'mantle time' to 'real time'. It meant the models should be run for 18 billion years, at present rates, to get the right number of overturns, rather than just the 4.5 billion-year age of the Earth. C&H had in any case only run their models for 3.6 billion years.

The resulting models easily yielded ages of 2 billion years or more. This was a notable result as the apparent isotopic ages had been taken by many to require mantle layering to keep mantle components separate for long periods. An example of the model results is shown in Fig. 14.1.

The models also yielded the right kind of difference between plate samples (from the top) and plume samples (from the bottom). They showed that most of the mantle (around 98%) had been processed through the melting zones under mid-ocean ridges, leaving very little 'primitive' mantle. That unprocessed mantle was scattered throughout the mantle rather than being concentrated in a deep layer, as many geochemists were still inclined to think.

A reviewer of my paper exploded that the latter result was clearly wrong, it was only plausible that around 30% of the mantle had been processed (as some other published models had found). He provoked me into finding a simple calculation to approximate the amount of processing, and this quite clearly confirmed my model results. That estimate was added to the revised version of the paper, thus improving it. It was also the basis for very useful further estimates that helped us to understand these and later models.

Models in which the tracers were negatively buoyant showed some of them accumulating at the bottom, in piles where the flow turned upwards, as had Christensen and Hofmann's models. This showed that settling to the bottom did not depend on there being a hot thermal boundary layer of lower viscosity, as they had concluded. These 'piles' were to assume significance as seismologists sharpened their mantle pictures and as models of mantle evolution improved.

Because I routinely had trouble with reviewers, who prolonged the process but often did not have objections that withstood analysis, and because the paper was about geochemistry, I decided to submit it to a primary geochemical journal (*Geochimica Cosmochimica Acta*), whose editor I knew, and knew would be scrupulously fair, to reviewers and to me. The paper was published in 2002 and the initial response was limited, partly because it did not match the usual journal audience. However over time it came to be fairly well cited.

To me the paper sorted out some important basic issues, but I suspect to many it just reworked what Christiansen and Hofmann had already done, perhaps tidying it up a little. The essential basic features of the modelling may not have been very visible to anyone who did not carefully read the paper.

This line of work was continued a few years later by my research associate Jinshui Huang. This was a rare time in my career when I was able to acquire some funding to support some extra help with the work. As I mentioned in the previous chapter, our funding arrangements had been changed. Whereas ANU used to get a block grant from the Federal government that was then filtered down through the hierarchy, we were, early in the new century, 'allowed' to apply to the national research funding body, the Australian Research Council, through which researchers in other universities were funded. The catch was that some of the ANU block grant was skimmed off and put into the ARC.

This was all in the name of competitive funding, the latest manifestation of the neoliberal ideology that competition always ensures the best result. It is a primitive conception of the world, and a perverse conception of people, who are highly social. Other universities were not so happy to have ANU people competing in the same pool, because many of us had little or no teaching responsibilities and stood a better chance of gaining funds. In due course that is what happened: ANU by and large got more back than was taken from it. Anyway by a minor miracle a proposal from me was funded, despite the widespread attitude that I was playing computer games and despite one of the most perverse evaluation processes I have had the misfortune to encounter.

Jinshui set about doing the same sort of modelling as I have just described, only in three dimensional models. This he did very capably, using computer programs generously supplied by John Baumgardner of Los Alamos and Peter Bunge of Munich and, later, Shijie Zhong of Colorado.

Broadly, the results confirmed the two-dimensional results: mean ages of simulated MORBs were about 1.8 billion years, mean ages of simulated OIBs were somewhat larger (I'll come back to that) and only around one percent of the mantle remained unprocessed through a melting zone.

However Jinshui made a key conceptual advance. Using my simple approximation of the rate of processing of material through the melting zone under a mid-ocean rise, that I had used to rebut a referee, I had defined a *processing time*, which is the average time it takes to process one mass of the mantle. Some mantle will remain unprocessed and other mantle might be processed twice or more, so this is the average. It depends simply on the average rate of seafloor spreading and the depth at which melting commences under a rise.

Jinshui was able to show why previous models, by us and others, had given a range of results for the average tracer age, from 2.7 billion years down to about 1.5 billion years. The reason was we had, inadvertently, used parameters that implied rather different processing times. It was mainly because we had assumed different depths of melting, or in some cases had limited the melting zone in other ways. Jinshui scaled all the results to their processing time and showed they gave consistent results, in both two and three dimensions.

Making the tracers heavier, thus simulating old oceanic crust that is denser than average mantle and tends to sink, increases the residence time of material likely to be sampled by a plume because it is at the bottom. This confirmed previous studies. We only evaluated the mean age of material at the bottom and did not simulate it being lifted by a plume. This probably over-estimated the age to be expected for OIBs because a plume probably would not pull up the oldest or densest material, which would tend to be in the lower middle of accumulated piles.

These studies clarified what the isotopic ages are really telling us. A common assumption at first was that they are the time in which chemical heterogeneities are removed by homogenisation—the coffee-stirring time or *homogenising time*. However melting also homogenises, in the sense of resetting the uranium-lead clock to zero. Then the other interpretation is that the lead isotopes are really telling of a previous melting event, on average 1.8 billion years ago. We can call this the *residence time*—the average time a sample is resident in the mantle between passages through a melting zone. The residence time would be similar to the processing time, except that the rate of processing changes with time because the mantle has been cooling, slowing down, and melting only at shallower depth. The processing time is a model parameter that controls the residence time, which corresponds to what we can observe.

The upshot is that the lead isotope ages can most successfully be interpreted as *residence times*, not homogenisation times, which are much longer in the whole mantle. And we had reached an understanding that the main control on the residence time is the processing time. Other factors, such as the viscosity structure of the mantle, the geometry (two- or three-dimensions, spherical or rectangular) or the strength of heating are secondary. The third of Jinshui's studies was done at full mantle heating,

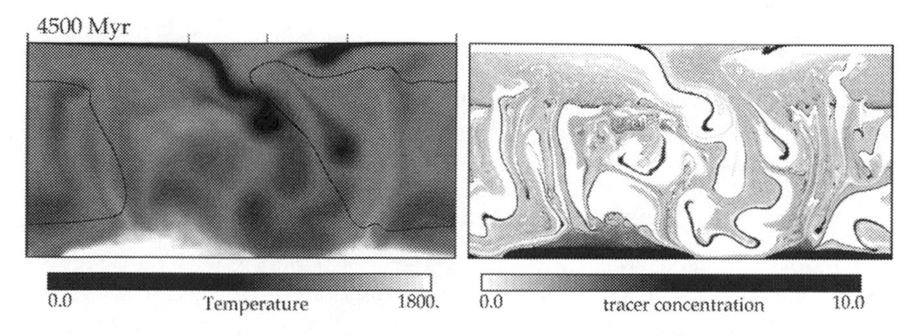

Fig. 14.2 Temperature **left** and tracer concentrations **right** at the end of an evolving mantle model. The tracers simulate oceanic crust that is subducted at two locations and sinks into the deeper mantle. Two substantial accumulations of tracers can be seen at the bottom. They are quite dense near the bottom and grade into normal concentrations higher up. The 'piles' are also hotter, because their material does not freely stir into the rest of the mantle. From Davies [37], reproduced with permission

by modelling only a region of the mantle on the latest computer. The results were still consistent.

A further development of this line of work was to start models with a higher temperature, as the mantle would have had early in Earth history, and let it evolve to the present, cooler and slower state as radioactive heating declined. This required some further development of the computer code and more computer power. Actually by this time there was more than enough power in the available computer, but it required special programming to take advantage of multiple processors in the computer. I did a less-challenging version of this reprogramming to use four processors instead of one.

New phenomena appeared in the hotter (early) mantle. The denser oceanic crust component tended to settle through the upper mantle and leave it moderately depleted, in spite of the convection being whole-mantle. The feature most pertinent to the present point is that large accumulations of oceanic crust developed at the bottom early on and these tended to persist into the present. The result was accumulations that were much larger than those formed by just running the model at present conditions, rather than cooling it from a hot condition. An example is shown in Fig. 14.2. Seismologists have detected two regions of the deep mantle that seem to be of this kind, one under Africa and the other under the Pacific. Other modellers have also found such accumulations, in both two-dimensional and three-dimensional models.

This study from 2008 also looked at the effect of a crustal density reversal between 650 and 750 kms depth, which might cause the mantle to convect in separate layers. Because of its different composition, the subducted oceanic crust component of a subducted slab does not transform to denser phases until 750 kms depth. This means it is positively buoyant between 650 and 750 kms instead of negatively buoyant. Uli Christensen in 1997 had tested the effect on subducted slabs in the present mantle and found that it could delay them at that depth but they would eventually sink through into the lower mantle. However in the longer term the oceanic crust component does not go away, it just gets dispersed around the mantle, as in Fig. 14.1. M. Ogawa had done several models that showed the crust material might accumulate at the transition zone and eventually block all flow. He called this the *basalt barrier* effect. However his models included other features, such as 'plates' that did not behave so much like observed plates, that made them not so easy to interpret.

I ran some evolutionary models that developed a repeating sequence of layering, due to the basalt barrier, followed by dramatic breakthroughs and then the re-establishment of the layering. After about 1.5 billion years the layering ceased, as the mantle cooled, and thereafter the behaviour was more continuous. So layering is a plausible possibility for the early history of the Earth, but still unlikely at present. The tectonic regime in the Archaean, before 2.5 billion years ago, has left a different style of imprint on continental crust surviving from those times, but whether that might be due in part so such episodic layering remains conjectural as far as I know.

This series of models, from Christensen and Hofmann in 1994 and running through the others just discussed, seemed to establish that much of the chemical heterogeneity in the mantle could be attributed to the formation, subduction and recycling of oceanic lithosphere, comprising basaltic oceanic crust and its complementary depleted source layer. The survival of such heterogeneities for around 1.8 billion years was established as quite plausible. The age indicated by the lead isotopes can be interpreted as the average time between passes through the mid-ocean rise melting zone. Al Hofmann, in a couple of survey papers in 1997 and 2003 showed the original Hofmann-White proposal continued to serve well, with some refinements, for the 'refractory' trace elements (i.e. those that are not volatile like helium) and their isotopes.

However serious puzzles remained. The basic issue was that we could not account for the full complement of some elements that should be in the Earth. This was the reason the layered mantle idea persisted. We seemed to need an extra reservoir in which would be found the missing neodymium and uranium (and other refractory trace elements) and the missing argon and some 'primitive' helium (and other noble gases).

The origin of the geochemical layered mantle went back to Wasserburg and DePaolo in 1979 (Fig. 10.2), and there were two arguments. The claim that some samples came from a 'primitive' source was soon dropped, though the idea of a primitive source persisted. The other argument was that the amount of neodymium in the continental crust was only about a third of what should be in the Earth. This could be accounted for if most of the neodymium in the upper mantle (about one third of the mass of the mantle) had been extracted into the continents and the lower mantle was still 'primitive'. Neodymium was seized upon at the time because there was (and is) good reason to think meteorites tell us how much should be in the Earth.

There never was any clear evidence for primitive material in the mantle, in the sense of material that had never been processed through a melt zone, or rather only once as it was erupted for the first time to the surface. All along, when people appealed to mixing between a 'depleted' source and a primitive source, the lead isotopes clearly showed the 'primitive' source had been processed, even if its neodymium signature happened to be close to the primitive value (Fig. 10.4). Hofmann and White made no appeal to a primitive source, and Al Hofmann in later years stressed the lack of

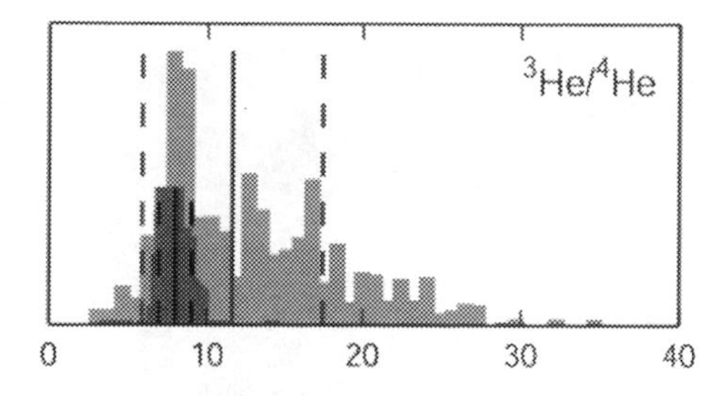

Fig. 14.3 Distribution of helium isotopic ratios from mid-ocean rise basalts (MORBs, **dark grey**) and ocean island basalts (OIBs, **light grey**). From Ito and Mahoney [24], reproduced with permission

evidence for a primitive source, but he also conceded he had no explanation for the 'missing' elements.

One geochemical signature that has been called 'primitive' is mischaracterised. Helium from some hotspots has a higher ratio of the two helium isotopes, $^3He/^4He$. ^4He is produced by the decay of uranium and some other radioactive elements, so it is 'radiogenic'. ^3He is not produced by any radioactive decay, so any that we find in the Earth has been here from the beginning. In that sense it is 'primitive', perhaps better called *primordial*, but so what? It says that not all of the helium originally present in the Earth has leaked out, but the amounts still detected are minute—a few thousand grams per year from the entire mid-ocean rise system. Its concentrations in rocks are measured in parts per billion (by weight) or less.

The $^3He/^4He$ ratio in mid-ocean rise basalts (MORBs) is around 8 times what it is in the atmosphere (Fig. 14.3; ^4He from the relatively radioactive continental crust has accumulated in the atmosphere). The ratios in some ocean-island basalts (OIBs) are more scattered but mostly they are higher, up to 30 or more. A few are less than 8. Those with high ratios were proclaimed to contain 'primitive' material that came from the primitive lower mantle. Granted there seems to be relatively more ^3He in most OIB sources than in MORB sources, but there is nothing says it is a primitive ratio because we don't know, to within orders of magnitude, how much helium was in the Earth at the beginning.

The $^3He/^4He$ ratio is important, but not because it is a signature of unprocessed mantle. We will encounter it again before long.

The other important argument about primitive or missing components involves argon, another noble gas. For argon there is another isotopic ratio, $^{40}Ar/^{36}Ar$. ^{40}Ar is produced from the radioactive decay of ^{40} K (potassium), whereas ^{36}Ar is primordial. We can't account for all the ^{40}Ar that ought to be in the Earth. Geochemists think they know how much potassium (K) is in the Earth, but will concede some uncertainties. Over the age of the Earth, the ^{40} K component of potassium will have been decaying into ^{40}Ar. Now helium is so light it escapes from the atmosphere, but argon is not so light and it does not escape. Therefore all the ^{40}Ar ever produced in the Earth ought to be around somewhere—in the atmosphere, the continental crust, the mantle or (possibly) the core. The core option would imply other chemical signatures that are not seen, so most geochemists allow there is probably very little in the core.

The atmosphere contains about half of the expected ^{40}Ar and the continental crust another 4%. A conventional estimate of the MORB source concentration by Don

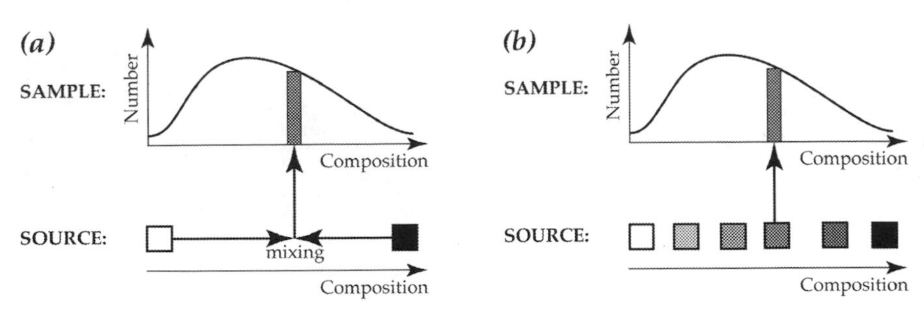

Fig. 14.4 Contrasting conceptions of mantle sources. **a** Generation of a sample of intermediate composition by mixing material from two *end-member* reservoirs before eruption. **b** Direct sampling of a heterogeneity of intermediate composition. From Davies [38], reproduced with permission

Porcelli and Gerry Wasserburg in 1995 implied only about 0.5% is in the upper mantle, so they presumed about 45% to be in the 'primitive' lower mantle. But if that MORB source estimate represents the whole mantle, which would then have about 1.6%, then we have accounted for only about 56% of the expected ^{40}Ar.

Claude Allègre was so enamoured of this argument that he gathered two other geochemical heavyweights, Al Hofmann and Keith O'Nions, as coauthors and published a short paper on it in 1996, though the argument was not original. Claude's approach to understanding the Earth was to beat opponents with a big stick until his version prevailed. I thought a more useful approach was to look at all the arguments to see if there might be uncertainties or holes in them.

As it happened I had found three reasons for wondering about the geochemical estimates. First, MORBs might be directly reflecting the heterogeneity of their source rather than a mixture of depleted and not-so-depleted reservoirs, and that would affect estimates of mean compositions. Second, my models had shown that even in a convecting mantle the uppermost mantle might contain less of the heavier, old subducted oceanic crust than the deeper mantle, so MORBs might underestimate some element abundances. Third, melting of a heterogeneous source would be more complicated than commonly accounted for, and might be less efficient at extracting incompatible elements into the oceanic crust.

Let us look first at the implication of sampling the heterogeneous mantle. Remember when Gerry Wasserburg laid out his interpretation of the mantle with two layers, one primitive (Chap. 10)? Not only did he propose the idea of layering, he planted the ideas that there are uniform reservoirs and that their compositions are beyond the extremes of the observed distribution. That interpretation is depicted in Fig. 14.4a. This way of conceiving of mantle sampling has dominated interpretations ever since, through the concept of *end members*. Hotspot basalts (OIBs) have been classified into several different kinds (EM-1, EM-2, HIMU, etc.; Fig. 10.4). They have been interpreted as being end members, along with DMM, the 'depleted MORB mantle', and the observed distributions as being mixtures of end members.

An alternative interpretation is shown in Fig. 14.4b. It is that an observed intermediate composition simply reflects its extraction from a source of intermediate composition. If the mantle is heterogeneous, which no-one seriously disputes, then its distribution of compositions will be reflected in the observed distribution of MORB and OIB compositions. There may be some local averaging, especially at mid-ocean rises where there are larger degrees of melting, because rising magmas tend to collect and mix in magma chambers before eruption. But any such averaging must be local and limited because there are large variations in neodymium isotopes, for example, even along the Mid-Atlantic Ridge.

This distinction matters. It means the MORB source, the real mantle tapped by mid-ocean rises, is not represented by the white box in these sketches, it is represented by the average of the distribution. Rather than being very depleted of incompatible elements (the depleted MORB mantle), the average MORB source is rather less depleted. It will contain more of the 'missing' elements than conventional estimates.

How much more was not so easy to estimate, because hardly anyone was thinking about average compositions, they were so stuck on the idea of end members. Charlie Langmuir, of Columbia and then Harvard, and his group had looked at the full archive of MORB compositions and extracted various averages and ranges. But the detailed data were in an unpublished thesis by Yongjun Su in 2002. I looked at the PetDB database (Petrology DataBase), which has a category 'All MOR'. An average of this yielded a uranium content of 240 ng/g (that's nanograms of uranium per gram of rock; geochemists commonly say 'ppb', meaning parts per billion, but that does not tell the uninitiated whether the ratio is by weight or mole or even volume, so I make the weight ratio explicit). But it was not clear what exactly was included in the All-MOR set without doing an exhaustive analysis. Kathleen Donnelly and others quoted two numbers in 2004, derived from the Su data set, an average for the East Pacific Rise of 80 ng/g and a particular E-MORB of 305 ng/g. More recently a published full analysis by Allison Gale, in 2013, has given an all-MORB average of 119 ng/g.

These numbers are all larger than a widely-quoted value of 47 ng/g from Shen-Su Sun and Bill McDonough in 1988. The indication is that average MORB is more enriched than 'normal MORB' estimates by a factor of 1.7–2.5, and possibly more if more 'enriched' samples ought to be included.

Higher concentrations in MORB imply higher concentrations in the MORB source, as the incompatible elements are concentrated into the melt by a factor of 10 or so. Previous estimates for the MORB source were 3–4.7 ng/g, whereas my rough accounting of all reservoirs, using the above numbers, suggested 10 ng/g. Using Gale's more recent average suggests 9 ng/g, still two to three times larger than previous estimates. Similar arguments apply to other 'incompatible' elements, so it was looking as though the conventional estimates of elemental abundances in the Earth needed serious revision, and that some of the 'missing' elements might be found hiding in plain sight—in the MORB source.

I won't say much here about the computer models (like that in Fig. 14.2) of tracer stirring and settling. They indicate not only that denser, subducted ocean crust would settle in piles at the bottom, but that some would tend to slowly 'rain out' of the upper mantle, which has a lower viscosity than the lower mantle. The result in the models is that the uppermost mantle is depleted of the denser component by 10–30%

compared with the deeper interior of the mantle. This would imply that the mantle interior is more enriched in old oceanic crust, and its complement of incompatible trace elements, by a further factor of 1.1–1.4. We can just bear in mind that the depletion of the mantle might be even less than the MORB data imply.

Petrology is perhaps the messiest of the geological sciences. It is the study of how rocks melt, which is complicated, and prone to being affected by quite minor components. We need to go into this topic a bit, because it relates the surface rocks, the MORBs, to the mantle underneath, the MORB source. Delving into this subject involved some arduous reading. Eventually some important general points seemed to emerge. I'll walk you slowly through them.

I need to explain a few key ideas. A pure compound, like ice, has a particular temperature at which it melts, in that case 0 °C. A mineral is a pure compound; examples would be olivine, quartz or felspar. A rock is (usually) a mixture of several minerals. The upper mantle is mainly composed of a rock called peridotite, which is a mixture of olivine, two different pyroxenes and possibly some spinel or garnet (you don't need to memorise all this, there won't be a test). A mixture of compounds melts over a *range* of temperatures. Peridotite begins to melt around 1100 °C (the *solidus*) but may not be fully melted until 1600 °C (the *liquidus*). Between those temperatures is it *partially* melted.

Melting under a mid-ocean rise usually proceeds to something like 10% partial melting. The first catch is that the melt has a different composition than the solid rock, because some minerals have melted more than others. The melt composition is usually some kind of basalt, which is why the oceanic crust is mainly basalt. The second catch is that the details can be affected by minor components of the rock, especially by tiny amounts of dissolved water, which is why melting occurs at subduction zones, even though the lithosphere is cooler there. The third, fourth and fifth catches are that it matters whether the melt drains out as it forms, the melt may pool and mix in magma chambers nearer the surface, the magma chamber may partially solidify, changing the composition of the melt, the melt may react with surrounding rock, … you get the general idea. It's messy.

If the upper mantle, the MORB source, were made of one kind of peridotite, then petrologists would be able to work backwards and infer the composition of the source. Early in the plate tectonics story people made that assumption and inferred that the MORB source had partially melted by 10–20%. This would mean the incompatible elements, those that much prefer being in the melt than in the dense, confining mantle minerals, would have been concentrated by a factor of 5 or 10 relative to their source.

However the heterogeneity of the mantle became acknowledged, and in that case the process, and the logic, is not so simple. People have used elaborate schemes to infer the composition of the MORB source. Two studies in 2004 and 2005 had inferred the concentration of most elements in the mantle starting with models of melting and then using ratios of elements or isotopes known independently. To get to estimates of important elements like uranium you have to work through about eight ratios, accumulating uncertainties as you go. Vincent Salters and Andreas Stracke started from a melting model to work backwards from MORB to the source content of *major elements*, whereas Rhea Workman and Stan Hart started from one mineral

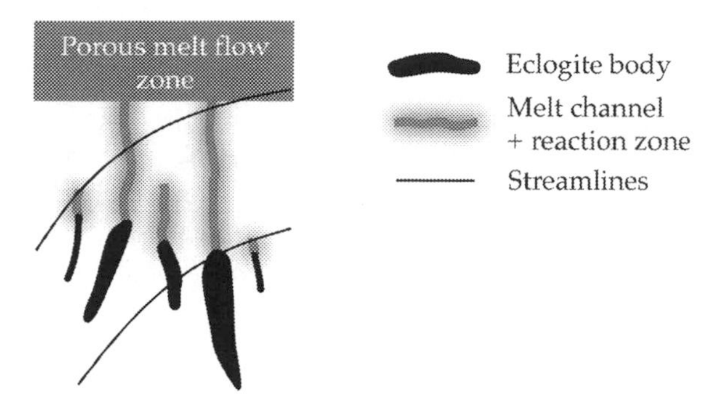

Fig. 14.5 Sketch of the migration of melt emerging from eclogite (or pyroxenite) bodies. The melt will react with surrounding peridotite, and will tend to refreeze. Some might react completely and become trapped in the mantle, whereas other melt might migrate a significant distance and reach the porous peridotite melting zone. Curves indicate streamlines of the mantle flow. After Davies [39], reproduced with permission

in a peridotite found at the sea floor and inferred the composition of the rock before melt was extracted. In each case it was plausible that the inferred source may not have equilibrated fully with minerals no longer present.

When we say the mantle is heterogeneous we mean there are parts that are made of different rock types than peridotite. Most commonly these are pyroxenite, comprising two kinds of pyroxene with other lesser components, or its higher-pressure form eclogite. This rock type corresponds roughly to subducted ocean crust, i.e. it is equivalent to a basaltic composition, though the minerals take different forms in the mantle environment. The pyroxenites, as I will call them, take the form of lumps and streaks that might be hundreds of metres in extent or thin streaks of less than a centimetre (there are places where slices of the mantle have been pushed up onto the land where they can be seen). Crucially, they will contain higher concentrations of incompatible elements than the peridotite, in fact they will probably contain most of the incompatible elements in the mantle.

Now, finally, to the point. As heterogeneous mantle rises under a mid-ocean rise the pyroxenite streaks will start melting before the surrounding peridotite. What happens then gets complicated. I know, even more complicated. Figure 14.5 is a sketch of the situation. If some of the melt from the pyroxenite (or eclogite, in the sketch) rises out of the pyroxenite (because it is less dense and can permeate through cracks) it will react with the peridotite it moves into, and it may re-solidify into a hybrid composition that I will call *hybrid pyroxenite.* If enough melt emerges, it may forge a path up through the mantle and reach the zone where the peridotite starts to melt, and thence to a magma chamber, or even the surface.

Mantle that rises directly under a rise crest will rise vertically until its peridotite also starts to melt. The melts from the two sources will end up merging and the end product will be as if the source were homogeneous. However mantle that rises *off-axis* will not rise all the way up, it will turn horizontally and move away with the new plate that has formed above it. Then the two kinds of melt will have only partially homogenised, or the pyroxenite may completely miss the peridotite melting zone. In the latter case the hybrid pyroxenite will not be erupted, it will stay in the mantle, and so will its complement of incompatible elements.

Thus, the final point, incompatible elements may not be extracted very efficiently from the mantle at mid-ocean rises, and a lot may remain in the mantle to circulate around for another billion years or two. The old presumption was that incompatible elements were efficiently stripped out of the melting zone, and over time the mantle would be steadily depleted of incompatible elements. The new picture is that this process is rather less efficient than had been supposed.

This picture also implies that the incompatible elements become concentrated *within* the mantle, in both the hybrid pyroxenite and the old MORB. Rather than the incompatible elements existing in a relatively uniform *depleted MORB mantle*, whose composition is taken to represent a large volume of the mantle, they exist in heterogeneities comprising perhaps ten percent of the mantle and need to be counted as depicted in Fig. 14.4.

The new story accentuates the question of how to relate the surface MORBs to the underlying mantle. If you tried to estimate the concentration of incompatible elements by working backwards from MORB you might under-estimate the amounts. The reason is they would not be extracted from as wide a melting zone, and as large a volume, as the apparent degree of melting might suggest. If they are from a narrower, smaller melting volume then they must have had a higher concentration. Thus the expected behaviour of melting in a heterogeneous source was generally consistent with the higher concentrations in *average* MORB than in the ill-defined *normal* MORB.

Around this time I reckoned I had a pretty good story, so I wrote a paper and submitted it. My thinking was running further ahead too, as we'll get to in a little while, and I reckoned I had a *really* good story. Referees had various comments and questions about the submitted paper, but it was the editor, Vincent Salters, who said I had not properly understood how concentrations in the mantle are estimated. He was lead author on one of the two relevant papers mentioned earlier, so he would know. I think at that stage I had implied they had used an estimate of the degree of partial melting, as described in the previous paragraph. He said they had avoided that assumption. My story of a less-depleted MORB source seemed to have collapsed. I was deflated.

This was difficult. Some important things had seemed to be reconciled in my story. It felt as though it *ought* to be true. But quite a few scientists have come unstuck believing their interpretation *ought* to be true. Well, I could have that feeling, but I couldn't base any firm story on it.

Eventually I determined to carefully examine how the estimates of mantle concentrations *had* been done. It turned out Salters and Stracke anchored their estimates to a peridotite (an anhydrous spinel peridotite, to be precise). I realised there was no assurance their peridotite reflected the mean composition of a heterogeneous (peridotite plus pyroxenite) source. In fact Salters and Henry Dick had reported a couple of years earlier that peridotites from the Indian Ocean could not explain the composition of nearby oceanic crust, whose trace elements required additional input from

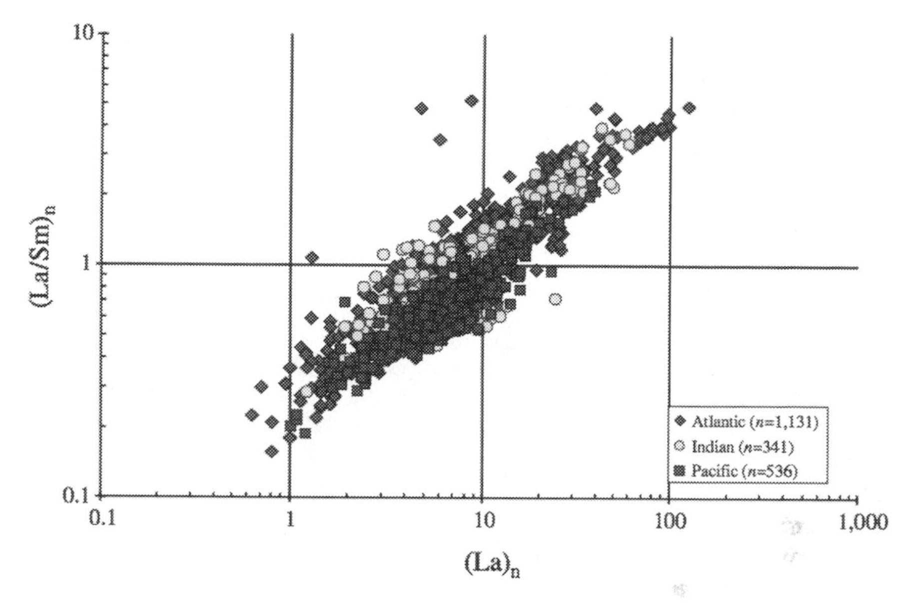

Fig. 14.6 Lanthanum/samarium versus lanthanum concentration, normalised to an estimate of primitive mantle. From Hofmann [40], reproduced with permission

something like a pyroxenite. The implication was that the more easily melted pyroxenites had been melted out of the source without their complement of incompatible trace elements being reflected in the residual peridotite.

In a similar way Workman and Hart had anchored their estimates to one mineral in a peridotite, which equally may have failed to register the presence of more enriched pyroxenites. Both studies were focussed on estimating the composition of the putative DMM, the *depleted MORB mantle*, so their estimates did apply to the more depleted end of the spectrum, but I wanted the average MORB source, and their estimates did not supply that.

Al Hofmann in 2003 had pointed out another problem with the concepts of *normal MORB* and its DMM source. I have already explained there seemed to be no clear way to separate normal MORB from not-normal MORB (Fig. 10.3). He showed the plot in Fig. 14.6 of La/Sm *versus* La (lanthanum and samarium, in case you're wondering, two trace elements). Normal MORB was often defined as having normalised La/Sm less than 1, but there is a continuous distribution across this value. Worse, noting the logarithmic scales of this plot, it is clear that much lanthanum would be excluded by this criterion, because the excluded samples have La contents up to ten times the cut-off value. If there was more lanthanum in the MORB source there would be more of the other incompatible elements as well.

So it seemed I had my story back. I revised the paper, and it was eventually published in 2009. It removed most of the need for a hidden reservoir in which to put the 'missing' components. The missing components were, in this story, distributed through the interior of the mantle. They had been overlooked largely because of a focus on the putative depleted end member, which was a hangover from the original layered mantle proposal. The depleted end member is there, it is also distributed around the mantle, another *source type* in the spectrum of source types. It tells us about a process, the extraction of magma, not about a putative reservoir.

Two significant questions remained. Although the amount of uranium indicated by meteorites could now be accommodated in the crust and mantle, that did not seem to be enough to keep the Earth cooking along—it implied the mantle is cooling faster than seems plausible. And there was still the question of where the noble gases might be hiding, especially the helium and argon.

In the meantime I was having further adventures with the promotions committee of the university. I was persuaded by colleagues to give it another go in 2005. In that year I had been awarded the Augustus Love Medal by the European Geosciences Union for my work on geodynamics, a significant gong. My book on economics, *Economia*, had been published the previous year and I included mention of it as 'service to the community' (a recognised, legitimate activity) and to note that publishing two books over the previous five years should be considered when assessing my output of papers, which was a little on the low side but nothing disastrous. Of course an assessment should in any case be about quality of work, not just number of papers per year. I was in another phase of developing computer code, and you get no credit for that.

The application was again unsuccessful. They were explicit in focussing on my recent record. They 'observed an apparent lack of sustained research contribution to your field in more recent years'. They chose, in the interview, to enquire what professional evaluations of *Economia* I could offer, though I had not presented it as part of my 'professional' output. Of course there were no such evaluations, the 'professional' academic economists (one of them sitting on the committee) would not deign to acknowledge the existence of someone questioning the legitimacy of the whole field of economics, as I explained as politely as I could in trying circumstances. (You are not allowed an advocate in these interviews.)

The guidelines for promotion were clear that the committee should make a whole-of-career assessment and that it should look for reasons to recognise excellence rather than reasons to exclude people. It was clear enough they had done neither. Of course it was all behind closed doors so that was impossible to prove.

The was a debrief interview in which the Professor focussed a lot on *Economia* (why?) and in which it was said 'not all referees were clear' that I met the criteria for leadership. There was mention of one referee saying Dr. Davies might have had a greater impact on the field if he had a different 'personality'. Oh, was that a criterion?

I later obtained referees' letters, redacted, under freedom of information rules. I found many very positive statements. I extracted key phrases that I thought to use in future applications or a possible appeal. I was going to include a small number here, but the full list is striking. It may be self-indulgent but here it is.

> without any doubts one of the leaders of this type of research . . . classic papers . . . a remarkable synthesis . . . impressively argued . . . fundamental contributions of major significance . . . the first researcher to develop models . . . a major accomplishment . . . [his] contribution is immense . . . 15 years ahead of his time . . . true milestones . . . one of the very best and

> most creative geodynamicists in the world . . . a world-renowned scientist of rare calibre . . . provides a very important service to society . . . a highly original scholar . . . exceptional contributions . . . one of the leaders in the field . . . a remarkable impact . . . an exceptional contribution to teaching in our field . . . a milestone . . . fundamental contribution . . . deep physical insight . . . a first-order contribution . . . an exceptional contribution to research in our field . . . [his] unique ability to cross disciplines . . . ample evidence of this remarkable trait . . . one of the leading workers in geodynamics worldwide . . . [his] work over the years has shaped a body of core knowledge in our field . . . a very important contribution . . . particularly insightful and effective . . . more than a powerful critique . . . a remarkable achievement . . . a 'breakthrough' contribution . . . a penetrating analysis . . . of great academic and practical significance . . . an exemplar of interdisciplinary work . . .

I would not have imagined such glowing tributes existed on the basis of any feedback from the Committee. Of course there were also constructive criticisms, but the overall tenor of the letters was highly positive. I was reassured about the attitude of my peers, it seemed there was not as much animosity as I had presumed. So why did the Committee decide that 'not all referees were clear'? That phrasing implies no referee actually said I did not meet the criteria, the Committee just made their own (negative) inference.

Was the Committee given reasons to recognise excellence? It would seem so. Did the committee instead look for reasons to exclude me, such as counting numbers of publications per year, *each* year. It would seem so: they 'observed an apparent lack of sustained research contribution to your field in more recent years'. Did they comment on the quality of those papers that did appear? No.

Why did they focus so much of their attention on *Economia*. I was clear the only reason I included it was to show that I had in fact been productive. They certainly seemed to find it a problem that I had done something not in my job description, and something they did not seem to know how to evaluate, except perhaps as a bit of irrelevant dabbling. Did they note, as I had, that perhaps it could be good to encourage senior people to do some adventurous work, perhaps risky (though not to them) but also perhaps with a big potential payoff? The ANU had been founded by people who thought that way.

One critical comment I found in the incomplete versions given to me was one referee's comment that I 'tend to be "self-centred" in my view of the profession and of RSES'. Gosh, no senior academic has ever been self-centred have they? Obviously some of the professional jealousy of the field found its way into referees' comments. So what? Does that diminish my accomplishments and intellectual leadership? Actually I found the comment ironic, because I tended to cite people in other 'camps' rather more than they cited me, as they tended to cite each other. Anyway a competent committee would be alert to professional jealousies. As an occasional journal editor I had found that divided referees' opinions could be a sign of innovative work.

There was no mention of the comment about my 'personality' in anything I was supplied with, unless it was that 'self-centred' remark. Again, what relevance would that have for the committee anyway? Perhaps it meant that I'm not naturally an alpha-male-type self-promoter and empire builder, but there are different kinds of leadership, including intellectual leadership, as I stressed whenever I was able. But it's true I had not built a little empire. They used the term 'academic leadership' and it was clear they meant turning out lots of students and postdocs. Einstein would not have found favour.

(It was not so easy for me to attract graduate students. Australian academia is a fairly small pond, the geology departments generally did not have anyone like me and the physics departments knew little about Earth science. There was a tendency for departments to hang on to their best students, which is a selfish attitude not in the best interests of students. Within my School the best applicants tended to be snaffled by senior people pulling rank. So I had not had many graduate students, and only a few postdoctoral associates.)

I think it's pretty clear that professional jealousies were at work both in my field and in the Committee. Especially early in my career I did not respond well to the acrimonious tenor of the field, which I have mentioned before. The field was divided into camps and I was not in any of the main ones, so that would have tended to discount my work. These are not valid factors for judging the quality of someone's work.

The Director of this time had told me that when he arrived he was 'warned' about me, so there was clearly some personal animosity around within the School. He also said he had never seen the person he was warned about. In other words he did not find me as unreasonable as come claimed, but I was certainly regarded as a trouble maker by others. I didn't just lie down and do as I was told. All my encounters with promotion committees were consistent with the view that I was regarded as a trouble maker. There is another factor involved here.

I had served a term as an academic representative on the University's governing Council during the 1990s. It was a tumultuous time in which the Government tried to split out the Medical School and then join the rest of the ANU with a nearby vocational university. The Vice Chancellor (the chief executive) of the day was caught out appeasing the Government for his greater personal power. There came a time in a Council meeting when the VC made a claim about what he had done that I knew to be untrue. As one of the more junior Council members I was the mug who spoke up and said that wasn't my understanding, as close as one comes to calling someone a liar. There was consternation. The VC was red faced. The meeting was adjourned for hurried discussions. It was papered over but it was probably another significant blow to the standing of a VC widely regarded as rather hopeless. There's more to the story but it shows I was indeed a trouble maker for some of the powers that wanted to be. I don't apologise for that of course.

Anyway I appealed the denial of promotion but of course it was denied in turn. It had been remarked somewhere in there that there was nothing to stop my Director paying me a Professor's salary. He promptly did that. Most of my immediate colleagues valued my work and were mystified by the failure to gain promotion.

I applied the next year, 2006, goodness knows why. My pitch included renewed (conventional) productivity and the need for a leading university to support adventurous thinking, a theme I had stressed all along. Denied again.

A colleague got his case reconsidered by pointing out his high productivity and very high citation rate, probably the highest in the School. I looked at my citation numbers and found I was well up among the Professors, and far beyond those in the rank in which I languished. I raised this in a meeting with a manager, but I did not get much reaction. They weren't very interested.

I was again resigned to my fate, but some colleagues still encouraged me to try again. I said I was not going begging my peers out there for yet another round of referees' letters. They were busy people and it was humiliating for me. Eventually I resolved to talk with the new Chair of the promotions committee. Could I submit another application without getting a new set of referees' letters? I would update all the other material but surely they had plenty of evidence on my performance. She allowed it might be possible, though I can't say I detected any enthusiasm.

So in 2008 I applied again, with no new referees' letters. Mine was the last interview of the day: clearly they did not want to disrupt others' interviews in case there was any trouble. I will note just one part of a very sorry experience. The first question was from a woman involved with education, not research. She asked me what teaching program I planned to institute as a newly minted Professor. It was a totally inappropriate question. Had she even looked at my file? I was a research academic and I had done rather more teaching than was required of me. I was obviously near the end of my career, not a newly minted junior teaching faculty member. Was she trying to catch me out? More likely she was just lazy and asked the same question of every candidate, I've seen people do that. In either case she was not doing her job. But it got me off to a bad start.

It was a very tense interview. There were other objectionable aspects that I won't go into. Someone asked how they could know whether my most recent work was any good if there were no new references. They were still going on as though my previous decade showed I was becoming lazy and stupid, and still failing to consider my whole record.

The interview had clearly gone very badly. I came out of it in a cold rage. I was due to meet with some friends. One of them said when I came into the room my face was white and it was obvious something major had happened. They let me unload. It helped greatly in those dire hours. They talked about 'the golden noose' of legal action, and cautioned against going that way. It would tie my life in knots for years, with no certainty of gain. It was good advice. They helped me to get my feet under me again, to steady and remember who I am and what I give to the world.

I retired from ANU in 2010, still with the rank I had come in at, 27 years earlier. A colleague along the way had once described being anything other than a Professor at ANU as 'perpetual adolescence'. Actually, my rank had been reduced, when they 'rationalised' it to be part of a consistent national system. In my judgement the university management had been as small-minded, defensive and risk averse as any box-ticking bureaucrats anywhere. They were not interested in the adventurous thinking the ANU's founders had promoted. They were and are a betrayal of what ANU, and all universities, ought to be. That's true of our society in general these days. An irony there is that much of that degradation is due to the economic ideology I was attempting to undermine. I should allow that the current Vice Chancellor is a scientist who seems to be genuinely interested in restoring ANU. I hope he can overcome some of the resistance to such notions that has grown over past decades.

Am I still resentful of that saga? Yes, but I don't lie awake at night over it. It comes up occasionally and I rage a bit and then it subsides. There are better things to do with life. Is this just a self-indulgent venting? You'll have to judge that for yourself. I think it demonstrates significant deficiencies in academic culture, and in an institution that still manages to do some world-leading work despite the hostility of ideologues and the limitations of small minds.

Chapter 15
Too Noble?

The geochemical picture of the mantle I had arrived at was that the MORB source is actually most of the mantle, as the geophysics indicated, that it is heterogeneous on all scales, as most people did not dispute, and that it contained two to three times as much of the incompatible trace elements as had previously been inferred. The heterogeneities would be of two main kinds. The first would comprise oceanic crust that had been subducted and stirred around for about 1.8 billion years on average. The second kind would be a range of compositions that I have called hybrid pyroxenite that would be produced when melt from old oceanic crust (MORB, in its eclogite form) reacted with the dominant peridotite component of the mantle. The incompatible elements would be concentrated in both the eclogite and the hybrid pyroxenite. Their higher overall abundances would remove most of the reasons for appealing to a hidden reservoir of less depleted or primitive material.

Still it was not clear how the noble gases might fit into this picture. There would, presumably, be more of them in the MORB source than previously thought, by the same logic as for the refractory trace elements: extraction should be inefficient and the extra gases would be left in the hybrid pyroxenite, to recirculate.

However the OIBs showed the distinctive feature that many of them had more of the primordial isotopes relative to the radiogenic isotopes, for example a higher ratio of $^{3}He/^{4}He$. Hofmann and White's proposal that old oceanic crust accumulated the D" region should mean there would be very little of the primordial isotopes, because the MORB should have been 'degassed' when it formed at the surface; any helium in D" should be mostly the radiogenic ^{4}He that had accumulated since. The weird thing about the OIBs was that the refractory trace elements fit the Hofmann and White story nicely, with clear evidence of having been processed through a melting zone, yet most of them had *un-radiogenic* noble gas signatures. So they seemed to be both processed and primitive at the same time.

All the clues were there, it required just one more insight, the realisation that came to me in the kitchen one night in 2009: *some* of the hybrid pyroxenite must be denser than the peridotite 'matrix' it is suspended in, just as the old oceanic crust is

G. F. Davies, *Stories from the Deep Earth*,
https://doi.org/10.1007/978-3-030-91359-5_15

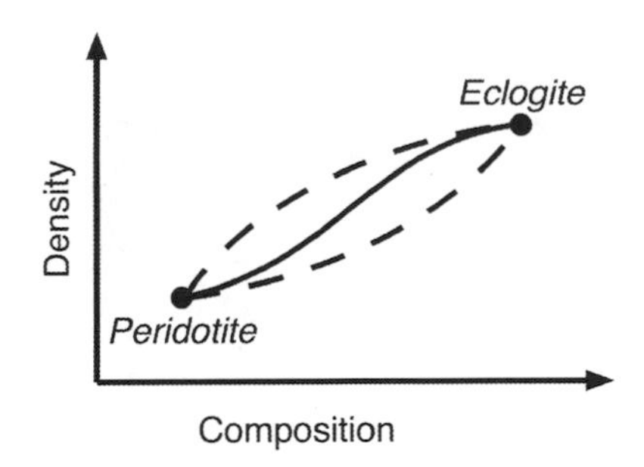

Fig. 15.1 Schematic of the possible variation of density between peridotite composition, comprising the bulk of the mantle, and eclogite, the form of oceanic crust in the upper mantle. The exact form of the variation is not known, but it is plausibly within the range of the dashed curves

denser (at most depths in the mantle). The variation of density with composition is illustrated schematically in Fig. 15.1.

Going through the compositional range, a hybrid that is made from mostly peridotite and only a little bit of pyroxenite would presumably have a density close to peridotite. At the other end, pyroxenite that is mostly eclogite with only a little bit of peridotite would presumably be denser than peridotite, like eclogite. Somewhere in between the density would grade from close to eclogite to close to peridotite. It might not be a simple proportionality, there might be an intermediate composition where the density did most of its changing. That doesn't matter. The important point is that some of the hybrid pyroxenite would be denser than peridotite. That means it would tend to settle to the bottom, just as some of the old oceanic crust settles to the bottom.

That means in turn that D" would hold a *mixture* of (degassed) old oceanic crust and (not fully degassed) hybrid pyroxenite. The reason for the conflicting signatures, processed but not fully degassed, would be that the two different components had quite different histories.

The key to the noble gases in OIB is that material accumulated in D" has a longer residence time than material floating around in the interior of the mantle, according to the models I and others had done, like that in Fig. 14.2. The residence time is the time between passages through a melting zone. In this interpretation, some of the melt is erupted at the surface where is it efficiently degassed, but some of the melt is not erupted, as proposed in Chapter 14. Thus passage through a melt zone removes some of the noble gases but not all, and the remainder recirculates within the mantle, some of it settling to the bottom. The sequence is sketched in Fig. 15.2. The mixture at the bottom may be pulled up in a plume, whose melting product will then reflect both the degassed oceanic crust and the less-degassed hybrid pyroxenite.

It turned out this was not the first time a mixture of sources in D" had been invoked. R. M. Ellam and F. M. Stuart at the Scottish Universities Environmental Research Centre had in 2004 reported correlations of helium isotopes with neodymium and strontium isotopes that led them to propose mixtures of a primordial component,

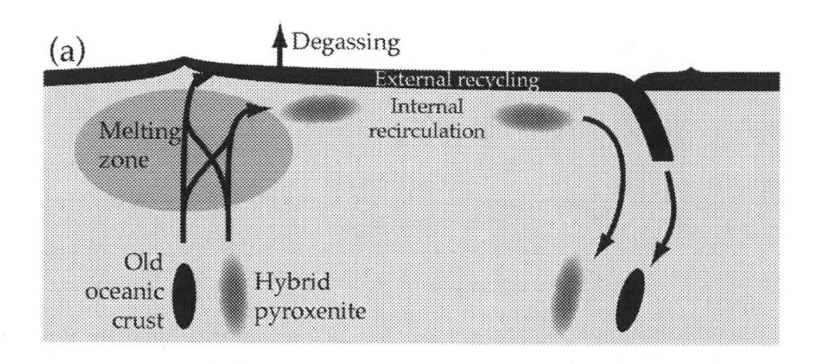

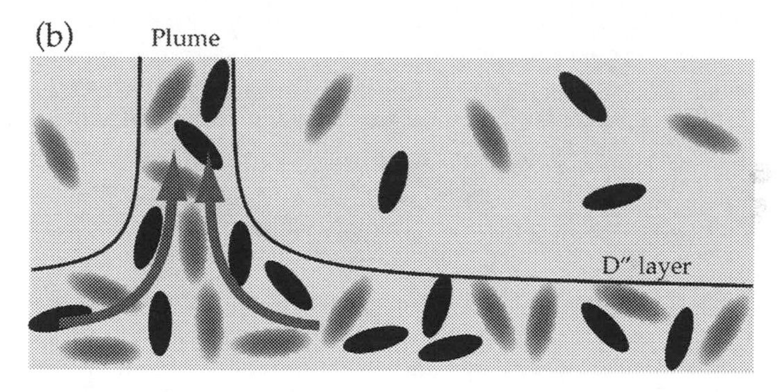

Fig. 15.2 Sketches of the fates of old oceanic crust and hybrid pyroxenite within the mantle. **a** The two kinds of heterogeneity rise into a mid-ocean rise melting zone, each melts, some of each melt is erupted and the rest recirculates without degassing. **b** The two kinds of heterogeneity float around the mantle, some accumulating at the bottom in the D" zone, where they may be pulled up by a plume. The plume source thus contains a mixture of degassed old oceanic crust and less-degassed hybrid pyroxenite. From Davies [41], reproduced with permission

containing more ^{3}He, with more usual depleted mantle component. They estimated only a few percent of a primordial component were required. They also argued for a minor component of material derived from continental crust. They noted that these components required only small mantle volumes, and a large, primitive lower mantle reservoir was not required. Cornelia Class and Steven Goldstein of Columbia University had proposed in 2005 that the plume sources with un-radiogenic helium had been isolated from the convecting mantle for 1–2 billion years, but did not require a large mantle volume. My story provided a straightforward mechanism by which these things could come about.

In my story the hybrid pyroxenite at the bottom has lost some gas, but not as much as that floating around in the mantle. The question then is whether this difference is enough to account for the differences between helium in OIBs *versus* MORBs, shown in Fig. 14.3. I realised it might be possible to calculate the different degassing histories using the same approach we had used to show there is little primitive mantle left. The strategy I used was to come up with a rough but plausible history of helium in the MORB source, then modify it according to the residence times for bottom accumulations in the models. You have to allow for some faster degassing early in Earth history—this had long-since been inferred from some isotopes produced by short-lived radioactive parents, and it was plausible because the initially-hot Earth would have cooled within the first few hundred million years until its temperature was sustained by internal radioactivity.

The short answer is that this approach worked nicely. The degassing of helium from the MORB source could be plausibly estimated, and it could reproduce estimates of present amounts in the MORB source. Then, simply by using a longer residence time, appropriate to the piles at the bottom, as guided by the earlier models, I got estimates for the OIB source that span the observed range (Fig. 14.3). Even better, it worked for neon and argon as well, though the present amount of argon came out on the low side. The argon does need some extra discussion, as we'll get to.

This seemed to be to be a major milestone. Whereas the noble gases had been a great puzzle, with no clear story of how they might fit into the mantle, this version was a consequence of the story developed to account for the abundance of the refractory trace elements, like uranium, in the MORB source, as told in the previous chapter. Not only that but it could be *quantified*, roughly but successfully. This was because the reason for the mixture was explicit and the *dynamics* could be estimated, going a step beyond the compatible proposals of Ellam and Stuart, and Class and Goldstein.

Perhaps it was possible to have only *one* model of the mantle, a model that could account for the physical observations, the dynamical modelling, the refractory trace elements and, now, the noble gases.

This was of course only the first cut, and more work refining the story would be needed. There were still a few outstanding questions. First, data for the heavy noble gases, krypton and xenon, were limited and it would remain to be seen if they were also consistent with the story. (As of this writing the xenon has been measured, and we'll get to that.) Second, there still seemed to be some 'missing argon', but the discrepancy was much smaller than in the conventional argument propounded by Allègre. Third, the amount of uranium in the Earth was still a question, as mentioned in the previous chapter. There were more developments to come on this topic too.

So I wrote a paper on this story, called *Noble gases in the dynamic mantle*, and it was published in *Geochemistry, Geophysics, Geosystems* in 2010. It was obviously on a much-debated topic and it was in a premier inter-disciplinary journal. There followed the deafening silence described at the beginning of this story. Nine years later it had been cited all of seven times. My better papers were usually cited at a rate of at least ten per year. The preceding paper on the refractory-element abundances, published in the same journal the previous year, had been cited 16 times in a decade.

I included the story in the last chapter of a new book, *Mantle Convection for Geologists* published in 2011. I wrote a long synthesis paper in 2011 on all the work Jinshui and I had done since my 2002 paper on mantle stirring; admittedly it appeared in a new journal *Solid Earth*, that was not mainstream (but they allowed me to use a lot of colour Figures). There was a session of an international conference in Melbourne in 2011, organised by colleagues to honour my retirement. Some prominent people were there and of course I described this work. The Goldschmidt international geochemistry conference was in Prague in 2011 and I spoke there, though only briefly. No response.

After all the sound and fury since the 1970s about the layered mantle, the primitive lower mantle, the sorting out of the roles of plates and plumes, the raging arguments with geochemists, the long development of the Hofmann-White version, the proclamations of the several *prima donnas* in each discipline, the silence was not only disappointing, it was deeply puzzling. It still is. I can suggest some factors, but they don't seem adequate.

Several of the prominent people left the field around that time. Perhaps no-one who was left cared very much. That does not seem very plausible, as there are still people writing papers about details of how the MORB source got depleted, for example. The art of measurement advanced enough to yield good data from the difficult case of xenon, which we will get to.

Probably many mantle geochemists did not like the claim that the previous estimates of trace element content were off by a factor of two or three. The scholarly response in that case is to cite the paper and give reasons why they may not be correct. It is not good scholarship to simply ignore a paper. Perhaps the conclusion was so at odds with previous perceptions that it induced a cognitive dissonance and people had to just shut it out. That seems a bit extreme.

Perhaps, as I was not a geochemist and not part of any of the several cliques then my contributions were not worth bothering with. But my 2002 paper on stirring had 76 citations and my 2008 paper on episodic layering, including accumulations at the bottom, had 35 citations—not exactly a best-seller but at least some notice of its existence. My more-cited papers had a few hundred citations over 25–30 years. I quote the numbers just to be clear how anomalous the response to the recent papers seems.

Perhaps the simple explanation is that people just did not understand the papers. The first involved some unfamiliar arguments, though the essence is clearly there and complicated argument is not unusual in petrology. As I noted earlier, first statements of a new insight are not always the most clear. The second paper, well I don't know why that would be so hard to understand, but a more recent development, which we'll get to, revealed that at least two people just failed to get the point. I can't say I think they were paying close attention though.

There was another story developing in parallel to these papers. It was that the Earth was not as much like meteorites as had been thought. This implied a shift in mean composition of the Earth that might account for some of the 'missing' components, though not all. Still, that story was by geochemists for geochemists and it got attention, and perhaps stole some of my thunder.

Perhaps I exited the field too soon. Perhaps I needed to spend a decade banging the drum, and the table, until people took it seriously. That didn't work for my argument about heat flow and topography though. My words seem to be suffering the fate of their subject, too aloof for the hoi-polloi to interact with.

There is a final twist, and un-twist, to my part of the story, along with the addition of some new observations and another option. The twist concerns another radiogenic isotope of neodymium, ^{142}Nd, produced by the decay of ^{146}Sm. It is less abundant than the ^{143}Nd we have encountered hitherto, and so harder to measure, but in a series of papers starting around 2005 Maude Boyet and Richard Carlson of the Carnegie Institution of Washington reported values for some key Earth rocks and meteorites. They found a distinct difference. The difference indicated that the average Earth is *not* the same as the so-called chondritic meteorites, as had been presumed for decades. Rather, the Earth seems to have somewhat lower abundances of the refractory trace elements we have been discussing.

This would mean there was less of the 'missing' components than in the conventional accounting and so less reason to appeal to a large primitive or less depleted reservoir. A lot of the story I have just been telling is about how to accommodate the amounts of trace elements that had been inferred from the meteorites, so if there was less to accommodate the issue was relieved. Relieved but not completely resolved, as there would still be some discrepancy, though not as large as had seemed before. In other words the depleted MORB mantle was not so depleted, relative to the new baseline, after all.

The story was reinforced when it was pointed out by G. Caro and B. Bourdon in 2010 that the least radiogenic helium from Baffin Bay, reported by Ellam and Stuart and discussed above, occurred in samples with the new average Earth compositions of strontium and neodymium. Perhaps those samples represented the real primitive Earth composition.

In my last synthesis in *Solid Earth* in 2011 I added the new twist to the story, allowing for both the old and new versions of average Earth. As the amounts of 'missing' components had been reduced, my version of where to find them was less urgently required. The new neodymium story seemed to steal a fair bit of my thunder. There was a simple geochemical story to account for much of the discrepancies, *versus* a complicated dynamical, petrological and geochemical story. Why bother with the hard story when a lot of the problem had gone away?

However in 2016 C. Burkhardt and others at Chicago, and independently A. Bouvier and Boyet, reported a more complicated story in which meteorites and Earth materials had been affected by nuclear reactions from solar neutrons very early in solar system history. They concluded that a difference between meteorites and the Earth could not be resolved after all. Several other studies followed confirming the more complicated story. The ^{142}Nd and some other isotopes were affected, but the overall composition was not changed. It is plausible that the Earth has an average composition similar to that traditionally inferred, though there may be a bit more uncertainty about that now. Anyway the Earth is not required to be different from the traditional composition.

In the meantime Sujoy Mukhopadhyay and associates at Harvard have been measuring xenon isotopes. This is an important advance, because there are several radiogenic isotopes of xenon that carry complementary information. Two of the isotopes are from parent isotopes that decayed away early in Earth history, and their daughter products tell us there was substantial degassing early on. However one of those daughters, ^{136}Xe, is relatively more abundant in OIB sources. This is consistent with the story developed for the light noble gases that the degassing process has been slower in the D" regions.

The other important result from xenon is to confirm that much of the xenon in the mantle has been recycled from the atmosphere. This had been indicated in earlier work by Ian McDougall and M. Honda at ANU, but Rita Parai and Mukhopadhyay in 2016 were able to constrain the proportion to 80–90%, in both MORB and OIB sources. Oceanic crust absorbs atmospheric xenon during reactions with sea water, so this result confirms that some of that xenon survives the subduction process and passes into the deep mantle, with a proportion settling into the D" zone. Thus the Hofmann-White hypothesis is confirmed fairly directly.

Several people have suggested variations on a theme that some material from the earliest Earth history has survived in D". In 2006 Igor Tolstikhin and others suggested that an early basaltic crust foundered and settled in the D" region, carrying with it primordial helium and other components. Really the D" region is capable of storing many things. The Tolstikhin story is specified in some detail, so in that sense it is rather conjectural, but it is not excluded from possibility. On the other hand a long-term, general process of subduction and settling, starting early in Earth history, is less conjectural and, it seems, capable of explaining our observations.

The revision of the ^{142}Nd story and the addition of the xenon observations struck me as opening the door to a paper pointing out their compatibility with the hybrid pyroxenite story. However there was a question. If hardly anyone had read the original noble gas story in 2010 I might need to review that work before pointing out how well it accommodates the new information. The trouble with that is a scientific paper is supposed to report original work. You are not supposed to go over old work unless the paper is specifically identified as a review. Of course you can summarise old work to set the context, but where is the balance? Another hazard was that I might not be aware of all the relevant work that had transpired in the decade since I had been active in the field.

I wrote a paper, including a fair amount of summary of the old work, and got the worst of both options. The paper was rather brusquely rejected by the editor as just a review of my own work, which it was not. On the other hand one reviewer, who had in the past been sympathetic to my ideas, wondered why I went on about hybrid pyroxenite, it just seemed a waste of space. So he completely failed to get the point. Another reviewer wondered why I invoke such an elaborate process to concentrate the helium when it would have been concentrated already in a primordial source. This reviewer evidently failed to appreciate that the generation of hybrid pyroxenite allows for the differential *preservation* of a small primordial signal in OIB relative to MORB. He also thought that hybrid pyroxenite that was derived from 'depleted mantle' would simply reflect its depletion, which is true but misses the whole story about a heterogeneous mantle being less depleted than previous estimates. Hence my conjecture above that perhaps people have simply not understood the story, so they have ignored it.

One reviewer liked the paper and recommended publication. Another tersely dismissed it. Several of the four reviewers pointed out work I had overlooked, which was fair enough. Such is the fate of many papers: some useful feedback, much misunderstanding, and conflicting recommendations. Other priorities deterred me from the major revision it would require.

We need to come back to the ^{40}Ar question. My estimates account for much of the total presumed to be in the Earth, but not for all of it. On the other hand the estimates of how much ought to be in the Earth are a bit indirect and subject to several uncertainties.

The ^{40}Ar abundance in the MORB source yielded by the evolution model described above is only about half of that required to balance the usual estimate of the Earth's ^{40}Ar budget. The inferred concentration in D" is a little higher than before, but D" and its related super piles are equivalent to only about 2% of the mass of the mantle, so it can hold only around 3%. With about 50% of the total budget in the atmosphere and 27% in the MORB source, this seems to leave about 20% still unaccounted for. However this deficit is rather smaller than previous estimates, which left about 45% unaccounted for.

^{40}Ar is the daughter of ^{40}K, a potassium isotope, and the estimated budget of ^{40}Ar requires an estimate of the amount of potassium in the Earth. This is not as simple as, for example, the budget of uranium, because potassium is relatively 'volatile', and some would have been lost from the Earth in the heat of Earth's formation. So the amount of potassium is usually estimated from the amount of uranium, which should be similar to its abundance in chondritic meteorites. However the K/U ratio of the mantle is debated. A ratio of about 13,000 is commonly used, but some argue it could be higher or lower.

Another issue is the amount of potassium in the continental crust. Ross Taylor and Scott McLennan in 1995 estimated the K_2O content of the continents to be 1.1%, whereas Roberta Rudnick and David Fountain in the same year estimated 1.9%. The continental crust is so heterogeneous it is doubtful any estimates have great accuracy. The average concentration of potassium in the crust and mantle could be as low as 203 μg/g, compared with the more usual estimate of around 240 μg/g.

This reduced potassium content (203 μg/g) would yield less ^{40}Ar over the age of the Earth. The evolution models described earlier yield between 90 and 100% of this amount. Thus a lower crustal abundance of potassium, and a consequent lower global abundance, is consistent with the present models.

Finally, we should not assume the Earth has retained all of its volatiles, because some of the atmosphere may have been blasted off by very large impacts during the late stages of the Earth's formation. Bombardment persisted until about 3.8 billion years ago, by which time about one third of the Earth's ^{40}Ar had been generated. If mantle degassing was fairly rapid during the first 0.5 billion years or so, as seems plausible, then it is quite conceivable that perhaps 10% or so of the Earth's ^{40}Ar was removed from the Earth entirely.

Thus the ^{40}Ar constraint is not as stringent as has been claimed. There are uncertainties in the amount of potassium in the continental crust and in the 'bulk silicate' Earth, and there may have been some early loss of ^{40}Ar from the atmosphere. The present mantle model may account for the total ^{40}Ar budget of the Earth when these uncertainties are allowed for. This is not a settled issue, because the points just made are vigorously debated, but we can say that at this stage the argon budget is not a decisive argument against the mantle story I have been telling.

Chapter 16
Perspective; Imperfect but Better Than Shouting

The story developed through this work integrates the main geophysical observations, fluid dynamical principles, diverse geochemical observations and a vast accumulation of geological observations into a single general model of how the Earth's interior works. That of course has been the goal of a great deal of work by many people, many more than have been mentioned here. There would still be many details to be explored and many aspects to refine, and that is a process with no definite end, but if the broad framework is reasonably accurate then those questions will be profitably pursued.

It is possible of course that the story will be found to be inadequate in some substantial way, such is the nature of scientific investigation and the contingent nature of the knowledge gained. My bet is that it will be borne out in broad terms, but I would say that, wouldn't I?

Charles Lyell might be pleased. We have concluded long-since, from field evidence, that mountains are pushed up slowly, but now we have a credible mechanism with good evidence behind it. When tectonic plates bring two continental masses together they pile up, because continental crust is too buoyant to be dragged down into the mantle; Fig. 8.7 is a sketch of the process. The plates are moving mainly because their own weight causes some of them to sink into the mantle, dragging the attached plate along behind and driving a circulation of the deep mantle.

Osmond Fisher, Arthur Holmes, David Griggs, Keith Runcorn and Harry Hess would probably also be pleased, because we eventually identified the moving plates as being part of a process of convection, unlikely as that seemed when Tuzo Wilson first outlined the plates for us. The plates are heavy because they are colder than the interior, therefore they are denser than the interior and inclined to sink, though their own strength, also due to being cold, may prevent them from sinking for a time. A plate is able to sink where the lithosphere is fractured, defining the contact between two plates. The cycle of mantle rising under a mid-ocean rise, turning horizontally, drifting and losing heat to the surface, sinking at a subduction zone and being reheated in the mantle interior is a process of convection. Convection is a heat transport mechanism driven by internal buoyancies (positive or negative) of a fluid.

G. F. Davies, *Stories from the Deep Earth*,
https://doi.org/10.1007/978-3-030-91359-5_16

Alfred Wegener would also be pleased, as his argument that continents have drifted has been thoroughly vindicated. His few advocates through the dark ages, especially Alex du Toit and Sam Carey, can also take a bow.

James Hutton, John Playfair and Charles Darwin might be pleased because a second mountain-building mechanism has been identified involving volcanoes. Hutton and Playfair argued for the role of heat and magma in geological processes, and we have identified volcanic hotspots produced by mantle plumes. Mantle plumes are part of a distinct second mode of mantle convection that takes heat from the Earth's core and transports it up through the mantle, though very little of the heat escapes the mantle directly, that is mainly the role of the plates. That plumes are an independent agent from the plates is not yet fully appreciated in the geological business, as it is still not uncommonly presumed that rising plumes are the 'return flow' of sinking plates. I hope I have explained clearly that that is not how plumes fit in.

Darwin would be pleased that his recognition of the age sequence of Pacific islands, from active volcano through eroded island to coral atoll, has been confirmed and was the clue that led Wilson and Jason Morgan to the idea of a relatively stationary mantle source, a plume. Darwin could also be pleased that his estimate of geological ages was much nearer the truth than Kelvin's, and that his perception that repeated earthquakes might raise mountain ranges has been well confirmed.

There is another class of volcanoes, associated with subduction zones. We have not discussed them much, as they are not central to this story, but they have important effects. They are one of the results of plate tectonics, so of a different origin than volcanic hotspots. They occur because oceanic crust reacts with sea water to form hydrated minerals, and those minerals break down under pressure as they are subducted, releasing the absorbed water. That water locally reduces the melting temperature of rocks, and the resulting magma rises and erupts along island arcs such as Indonesia or volcanically active margins such as Chile. They are thought to be part of the main mechanism that has distilled the continental crust from the mantle.

Charles Lyell might be less pleased that the island arc volcanoes are prone to violent explosions that occasionally have been catastrophically large. He might also be discomfited that large meteorites have occasionally caused planetary mayhem, and that flood basalt eruptions due to the arrival of a new plume 'head' can also cause global disruption. That there are occasional catastrophes does not detract from Lyell's great insight that most of the time geology is driven by inexorable creeping forces acting over immense spans of time.

Paul Gast, who pioneered the study of lead and strontium isotopes in rocks derived from the mantle, would presumably be pleased that we now have a plethora of isotopic systems that have given us a key timescale of mantle processing and that are used to trace the many detailed processes and products of mantle convection. That the study of isotopes and trace elements led to raging arguments probably would not surprise Gast, for that is the nature of science conducted by passionate human beings, though he might agree that a little less dogmatism would be more conducive to progressing our understanding.

Some isotopes are now used to trace the very earliest history of the Earth, giving us glimpses of 'a vestige of a beginning', in Hutton's phrase. Hutton's conception of a steady-state Earth building mountains, eroding them and building more over an indefinite time period is not the picture we have now, but it is not a bad approximation of the second half of Earth's long history.

When I came into this business the Pre-Cambrian, everything prior to 540 million years ago, was only known in sketchy outline, but by now far more detail is available, thanks to many kinds of radiometric dating (pioneered by Arthur Holmes early in the twentieth century), along with much field work and geochemistry. It is clear that things worked rather differently in the Archaean era, prior to about 2.5 billion years ago, because the patterns imprinted on continental crust surviving from then are rather different from the later pattern of long, roughly linear mobile belts. There is a lively debate about whether plate tectonics only started after the Archean, or whether it operated but in a significantly different way in earlier times.

The early geologic history of the Earth is another big subject, beyond the scope of this story. The questions are unlikely to be resolved as cleanly as the recent history because the evidence is far more fragmentary. Even so, it is amazing to me how specific our debates have become over half a century of investigation.

There are of course continuing lively debates about the present mantle. I have only briefly mentioned the 'super-piles' under Africa and the Pacific, known more dispassionately as LLSVPs: Large, Low Shear Velocity Provinces. They are sometimes called 'super plumes', but they are nothing like plumes. Seismologists have been able to define their locations and shapes in some detail. There is a widespread perception that they are compositionally different, probably a little denser than normal mantle and also likely to be warmer, so they would be close to neutrally buoyant. It is commonly allowed they may be related to the D" zone, perhaps being swept-up piles of similar material as I have portrayed them here. They may contain more primordial material. There is a distinct tendency of plumes to have started around their margins, which makes some fluid-dynamical sense. To what extent they comprise primordial material versus material generated more continuously through Earth history is a fairly open question. The nature of the D" zone itself is still the subject of detailed investigations and debates.

There is a debate about the amount of heat coming out of the core into the bottom of the mantle. Diverse lines of evidence are pertinent—the composition of the core, the thermal conductivity of the core, the history of the magnetic field recorded in ancient rocks, the history of cooling of the mantle, the strength of mantle plumes and whether plumes lose buoyancy as they rise through the mantle, for example. It is believed the solid inner core is crystallising out of the core as it cools very slowly, and this can induce a second form of core convection that might yield a magnetic field different in significant ways. I concluded in 2007 that a plausible resolution of apparently conflicting but rather uncertain evidence favoured a core heat flow on the low side of estimates, but this is a contested topic that could shift our understanding of the details of mantle dynamics in significant ways, though I would expect the broad picture I have painted to survive.

The amounts of trace elements in the mantle are debated. The main issues have been covered in the story I have told. The inventory of argon is still in question, though I noted several uncertainties that might yield a resolution. Doubtless there will be other puzzles too, such as whether the stories for D" (Hofmann and White's and mine) would result in too much radiogenic osmium in OIBs—this question was raised by a reviewer of my recent unpublished paper.

A larger outstanding question is how much uranium is in the Earth (and thorium and potassium, the other main heat generators), and how much of that is distributed through the mantle. I argued for there being more in the interior of the mantle than previous geochemical estimates, but even that is not enough to prevent the mantle cooling faster than it seems to have over the past few hundred million years (the total being taken from chondritic meteorites). It is possible mantle convection and plate tectonics proceed more variably or episodically than we at first assumed.

Some people's view is that the big question, the *grand challenge*, is why there are plates at all. In this view we will not be able to claim we understand the system until we can make models that determine and drive themselves, without prescribing where plate boundaries are or how fast the plates move.

In all of my models I have prescribed the locations of plate boundaries, and in most of them I have also prescribed their speed, though chosen to be close to the speed of a freely convecting fluid. This leaves my assumptions explicit and visible. It also makes for a cleaner 'numerical experiment' when, for example, the goal is to see how the plate-scale flow stirs heterogeneities. Some other modellers commonly have done the same. Others again argue this is unsatisfactory, not only because we cannot be sure what the appropriate velocity ought to be but more fundamentally because we should let the model find its own speed.

I take the point as a tactical issue, but I don't agree that self-determining models would be a fundamental advance, for two reasons. First, we don't know the details of the complicated nonlinear rheology that determines the resisting stresses, particularly within subduction zones. We may make a model that reproduces observed plate speeds, but we simply would not know if the rheology used in such a model is the same as in the Earth. The model in Fig. 11.3 determines its own speed, but I make no pretence that the lowered viscosity at the plate boundaries accurately represents real plate boundaries.

Rheology, the relationship between force and deformation, is clearly very complicated in rocks, especially at the intermediate temperatures and pressures relevant to subduction zones. It is also known to be affected by minor components, especially water. We may, through continuing study gain a general understanding of the rheology of relevant compositions, but we are unlikely ever to know the details in a particular subduction zone. In this circumstance choosing a complicated rheology for a model is really just hiding our ignorance deeper inside the model. I would rather keep my assumptions clearly visible, so people know they are there and they can change them if they want.

The second reason self-determining models are not very useful is that the location of plate boundaries is historically contingent. Plates break where past activity has left them weak. That past activity depends in turn on previous activity. It is not conceivable that we could start a model from two or four billion years ago and have it reproduce the present configuration of the plates. Not only will we never know in enough detail how to start the model, but such problems are intrinsically unpredictable in detail because very small changes can deflect their trajectories into completely different development. Mantle convection itself is close to chaotic in this sense, even without the vagaries of the plates at the surface.

Social and political aspects of science have played significant roles in these stories.

Competitive funding of basic research inhibits creativity, as it encourages people to stick with more routine work that involves less risk to continued funding. It is a political choice to fund science this way, and politicians typically have little understanding of how science is really done, and how the big pay-offs come to pass. It was expressed to me once in the US that Congress has a procurement mentality—they understand how to procure an aircraft carrier, and just think of buying scientific discovery in the same way. So we are supposed to propose what we will discover next and ask for money to discover it. As I have often remarked, if I knew what I would discover in five years' time I would not be doing my job.

Canada used to have a system in which you applied for continued funding based on your past three years' work, rather than what you might do in the next few years. This is far more sensible. I don't know if they still have this system.

Funding was different for a couple of decades after World War II. Many politicians and bureaucrats had come to understand the crucial role of scientific and technological innovation in winning the war. In the US much fundamental research was funded by the Office of Naval Research, even abstruse theory from general relativity. In Australia the ANU was set up to do elite research and given a generous block grant to spend as it saw fit. Some of the best expatriate Australians and others were hired, and trusted to do good work, which they did. Those days ended about the time I went to the US, and the disease reached Australia a little later.

Mistrustful bean counters and economists operated on the assumption that we were all lazy and just looking to feather our own nests, so we had to be made *accountable*. We had to have our *efficiency* and *effectiveness* rated. The fact that ANU regularly reviewed the performance of its various parts apparently did not count. Earth Sciences were rated by external reviewers as 'among the top five in the world in what they do', but we still had to detail how we spent our time and justify our next allocation of resources (generous or miserly according to standing). The fact that most of us worked far more hours than our contracts specified, and continued to make noted contributions, did not seem to be noticed.

Personalities have appeared in these stories, particularly the big bombastic or bullying ones, though the sly manipulative ones are there as well, if not much remarked. Despite their hubris, the ideas of Dan McKenzie, Gerry Wasserburg, Claude Allègre, Ted Ringwood and (to a lesser extent) Don Turcotte have not prevailed. The geochemists were very good at measuring, but the lasting interpretations often come from elsewhere, as Tuzo Wilson's example demonstrated. I will grant that at least Dan McKenzie did not continue to insist (publicly) on his version of mantle convection, as some others continued to insist on their favourite pictures, including his predecessor Harold Jeffreys. Dan at his best is also very smart, having given us some early clever analyses of subduction zones and the synthesis of sea floor spreading in the Indian Ocean that I noted earlier.

Modern academia is not so good at synthesis. It is often remarked that we work in silos, with disciplines isolated from each other physically, administratively and often intellectually as well. Particularly in this time of planetary crisis it is crucial that we take a holistic view of our situation, and there are many who understand that, but many resist it. I have in recent years seen an excellent proposal for very broad collaboration across disciplines, to address the planetary crisis from a systems perspective that includes social, physical and natural sciences, fail to attract support within ANU.

In the present story, it was essential to consider the geophysics and the geochemistry together, as both clearly had important things to say about the mantle, but most scientists do not cross that boundary, or even stray far from their sub-discipline. This makes it difficult to get ideas published and appreciated, as people tend to be less receptive to novelties from outside their field (unless delivered by a widely 'respected' authority, but of course we are supposed to judge according to the evidence, not authority).

Evidence, conjecture, confident assertion, hubris and the formation of cliques have all contended in the stories I have told. As a rather un-confident kid from the bush, a colonial even, this made it harder for me to gain traction. More generally the imperfection of the social processes of science, the involvement of factors other than clear exposition and assembled evidence, inhibits science to a degree.

Yet these social factors are present in everything else we human beings do. Science is not different in kind, in that respect. However it is different in degree. Journalists may claim to be sceptical and to have a bullshit filter, but it is commonly a very selective one with some gaping holes. Politics has even less of a bullshit filter, increasingly so these days, and we will all know an angry uncle character who will bang on loudly about his immovable convictions about politics or society.

The difference in science is that the bullshit filter is much better than in many other aspects of our lives. Quite a lot of bullshit still gets through, but over time most of it is filtered out. There *is* a lot of calm assessment and discussion. People *do* stop and look again at their arguments and the evidence (perhaps after a bit of groaning or complaining to colleagues to vent some frustration).

Alfred Wegener did not live to see his ideas widely accepted, but they were, and decisively, though it took several decades for new people and new evidence to be brought to bear. Charles Lyell, on the other hand, succeeded in having his ideas widely accepted within his lifetime, so much so that some people later had to face down fierce opposition to the idea that some geological processes can be very rapid.

So we have progressed from the catastrophism presumed by Europeans in order to fit their perceptions of great upheavals into a Biblical timescale. We have accepted the idea that continents wander over the Earth's surface, and that although the mantle is solid it behaves like a fluid over geological timescales. We have found a way to reconcile the brittle behaviour of the solid lithosphere with the fluid convection that many suspected had to be involved. And we have discovered much else besides, about the operations of the deep Earth.

We have arrived at a remarkable picture of the Earth. The planet beneath our feet is very slowly turning itself inside out. Our human experience of this is mostly indirect, through earthquakes and volcanoes, because the extrusion of new insides and the consumption of old outsides occurs mostly under the oceans. That is part of why it took so long to recognise what is happening.

It is easy to talk about large rigid plates, but they are remarkable in themselves. The Pacific plate is 14,000 kms across, which is about a third of the Earth's circumference. It is really an inverted bowl, a large spherical cap. As it is no more than about 100 kms thick it is remarkable that it retains its coherence, especially as it is pulled from widely separated places on its rather ragged western edge, notably around Japan in the north and from Tonga to New Zealand in the south.

We have not found any comparable behaviour among the solar system's planets and satellites. The Moon and Mercury have stable, unbroken lithospheres. Mars shows signs of early activity and perhaps an episode of overturn, but has been inactive for a long time. Venus seems to be active under its dense shroud of clouds, but the dominant pattern is of various kinds of local eruption and extrusion. With a surface temperature of several hundred degrees its lithosphere is probably too thin and weak to form large rigid pieces. It seems to have undergone a near-total volcanic resurfacing a few hundred million years ago, but it would be more like a giant flood basalt eruption than plate tectonics. It is possible the Earth underwent such overturns during the Archaean, but that is conjectural. Some of the larger icy and rocky satellites of the giant planets show signs of past or present activity, but nothing like plate tectonics is evident.

It may be that the occurrence of plate tectonics reflects a delicate balance between internal forces and a lithosphere weak enough to be broken but strong enough still to exist as large pieces. Computer modelling by a number of investigators tends to bear this out, with three or more regimes ranging from unbroken lithosphere through fragmented lithosphere to a thin and deformable boundary layer that we would not call a lithosphere.

There are interactions between the interior and the biosphere. The biosphere affects rates of weathering and erosion and the chemical composition of sediments that are deposited on the sea floor, some of which are subducted. Volcanism generated by plates and plumes brings nutrients to the surface and perhaps is important to the maintenance and proliferation of life.

Some have claimed that without life there would be less water or other minor components subducted and the mantle would be too stiff to flow and allow plate tectonics. However this overlooks a self-regulating mechanism of the mantle: if the mantle cannot remove its heat its temperature will rise and its viscosity will drop until it becomes mobile enough to remove heat by one means or another.

Earth is unique in our limited sample of well-observed planets and satellites. It also seems to be rather unusual among the thousands of exoplanets that have now been detected orbiting other stars. Smaller planets close enough to their star to have temperate surfaces tend still to be somewhat larger than Earth, five to ten times Earth's mass. The depths of the mantles of such ‘super-Earths’ would be hotter because of the greater pressure, and this might lead to extremely hot and violent eruptions that would make life difficult on the surface.

Though we now know that planets are very common in the galaxy, it may be that Earth is still highly unusual because of a combination of particular circumstances. Having the surface temperature vary by no more than a few tens of degrees over billions of years is certainly a remarkable feature, evidently due in part to self-regulating mechanisms according to James Lovelock’s Gaia hypothesis. Avoiding the complete destruction of life from without and within also seems to involve some good fortune.

For any number of reasons still only poorly understood, it seems that Earth is unusual, and unusually hospitable. It is still true there is no Planet B. We need to cherish our beautiful home in a hostile universe.

Bibliography of Sources of Figures

1. Ricketts B. It only takes a moment; the ups and downs of earthquakes. Geological digressions. 2017. https://www.geological-digressions.com/it-only-takes-a-moment-the-ups-and-downs-of-earthquakes/.
2. Isacks B, Oliver J, Sykes LR. Seismology and the new global tectonics. Jour Geophys Res. 1968;73:5855–99.
3. Benioff H. Seismic evidence for crustal structure and tectonic activity. Geol Soc Amer Spec Paper. 1955;62:61–74.
4. Davies GF. Dynamic Earth: Plates, Plumes and Mantle Convection, 460 pp., Cambridge: Cambridge University Press; 1999.
5. Wegener A. The origin of continents and oceans. Translated from the 4th revised German edition of 1929 ed. London: Methuen; 1966.
6. Mason RG. A magnetic survey off the west coast of the United States. Geophys J Roy Astr Soc. 1958;1:320–9.
7. Vine FJ. Spreading of the ocean floor: new evidence. Science. 1966;154:1405–15.
8. Atwater T, Severinghaus J. Tectonic maps of the northeast Pacific, in The Geology of North America, Vol. N, The Eastern Pacific Ocean Hawaii; Winterer EL, Hussong DM, Decker RW, Editors. Geological Society of America: Boulder, CO.; 1989.
9. Holmes A. Continental drift: a review. Nature. 1928;122:431–3.
10. Wilson JT. A new class of faults and their bearing on continental drift. Nature. 1965;207:343–7.
11. Lithgow-Bertelloni C, Richards MA. The dynamics of cenozoic and mesozoic plate motions. Rev Geophys. 1998;36:27–78.
12. Maxwell AE, et al. Deep sea drilling in the South Atlantic. Science. 1970;168:1047–59.
13. Heirtzler JR, Le Pichon X, Baron JG. Magnetic anomalies over the Reykjanes ridge. Deep Sea Res. 1966;13:427–43.
14. Dewey JF, Bird J. Mountain belts and the new global tectonics. J Geophys Res. 1970;75:2625–47.
15. Atwater T. Plate tectonic history of the northeast Pacific and western North America, in The Geology of North America, Vol. N, The Eastern Pacific Ocean and Hawaii. Winterer EL, Hussong DM, Decker RW, Editors. Geological Society of America: Boulder, CO.; 1989.
16. Woods MT, Davies GF. Late cretaceous genesis of the Kula plate. Earth Planet Sci Lett. 1982;58:161–6.
17. Isacks B, Molnar P. Distribution of stresses in the descending lithosphere from a global survey of focal-mechanism solutions of mantle earthquakes. Rev Geophys Space Phys. 1971;9:103–74.
18. Davies GF, Richards MA. Mantle convection. J Geol. 1992;100:151–206.
19. Davies GF. Whole mantle convection and plate tectonics. Geophys Jour Roy Astr Soc. 1977;49:459–86.

G. F. Davies, *Stories from the Deep Earth*,
https://doi.org/10.1007/978-3-030-91359-5

20. Davies GF. Viscous mantle flow under moving lithospheric plates and under subduction zones. Geophys J Roy Astr Soc. 1977;49:557–63.
21. Lux RA, Davies GF, Thomas JH. Moving lithospheric plates and mantle covection. Geophys J Roy Astr Soc. 1979;58:209–28.
22. DePaolo DJ, Wasserburg GJ. Inferences about mantle sources and mantle structure from variations of $^{143}Nd/^{144}Nd$. Geophys Res Lett. 1976;3:743–6.
23. Wasserburg GJ, DePaolo DJ. Models of earth structure inferred from neodymium and strontium isotopic abundances. Proc Natl Acad Sci USA. 1979;76:3594–8.
24. Ito G, Mahoney JJ. Melting a high 3He/4He source in a heterogeneous mantle. Geochem Geophys Geosyst. 2006;**7**. doi:https://doi.org/10.1029/2005GC001158.
25. Hofmann AW. Mantle chemistry: the message from oceanic volcanism. Nature. 1997;385:219–29.
26. Hofmann AW, White WM. Mantle plumes from ancient oceanic crust. Earth Planet Sci Lett. 1982;57:421–36.
27. Davies GF. Ocean bathymetry and mantle convection, 1. Large-scale flow and hotspots. Jour Geophys Res. 1988;93:10467–80.
28. ETOPO5. (Topography of the Earth, 5' grid). National Geophysical Data Center, U.S. National Oceanic and Atmospheric Administration, 325 Broadway, Boulder CO; pp. 80303–3328.
29. Marty JC, Cazenave A. Regional variations in subsidence rate of oceanic plates: a global analysis. Earth Planet Sci Let. 1989;94:301–15.
30. Sclater JG, Jaupart C, Galson D. The heat flow through the oceanic and continental crust and the heat loss of the earth. Rev Geophys. 1980;18:269–312.
31. Davies GF. Effect of a low viscosity layer on long-wavlength topography, upper mantle case. Geophys Res Lett. 1989;16:625–8.
32. Schoene B, et al. U-Pb geochronology of the Deccan Traps and relation to the end-Cretaceous mass extinction. Science. 2015;347:182–4.
33. Richards MA, Duncan RA, Courtillot VE. Flood basalts and hot-spot tracks: plume heads and tails. Science. 1989;246:103–7.
34. Griffiths RW, Campbell IH. Stirring and structure in mantle plumes. Earth Planet Sci Lett. 1990;99:66–78.
35. Davies GF. Mantle plumes, mantle stirring and hotspot chemistry. Earth Planet Sci Lett. 1990;99:94–109.
36. Davies GF. Stirring geochemistry in mantle convection models with stiff plates and slabs. Geochim Cosmochim Acta. 2002;66:3125–42.
37. Davies GF. Episodic layering of the early mantle by the 'basalt barrier' mechanism. Earth Planet Sci Lett. 2008;275:382–92.
38. Davies GF. Mantle convection for geologists, 240 pp., Cambridge: Cambridge University Press; 2011.
39. Davies GF. Reconciling the geophysical and geochemical mantles: plume flows heterogeneities and disequilibrium. Geochem Geophys Geosyst. 2009; 10:19 https://doi.org/10.1029/2009GC002634
40. Hofmann AW. Sampling mantle heterogeneity through oceanic basalts: isotopes and trace elements. In: Carlson RW, editor. Treatise on Geochemistry Vol 2: The Mantle and Core. Oxford: Elsevier-Pergamon; 2003. p. 1–44.
41. Davies GF. Noble gases in the dynamic mantle. Geochem Geophys Geosyst. 2010;11:Q03005. https://doi.org/10.1029/2009GC002801.

Printed in the United States
by Baker & Taylor Publisher Services